Poetics of the Earth

Poetics of the Earth is a work of environmental philosophy, based on a synthesis of eastern and western thought on natural and human history. It draws on recent biological research to show how the processes of evolution and history both function according to the same principles.

Augustin Berque rejects the separation of nature and culture which he believes lies at the root of the environmental crisis. This book proposes a three stage process of "re-worlding" (moving away from the individualized self to become a part of the common world), "re-concretizing" (understanding the meaning and historical development of words and things) and "re-engaging" (reconsidering the relationship between history and subjectivity at every level of being) in order to bring western thought on nature and culture into sustainable harmony and alignment.

This book will be of great interest to students and scholars of environmental studies, environmental philosophy, Asian studies and the natural sciences.

Augustin Berque has recently retired as Director of Studies in Environmental Philosophy and Geography at the École des hautes études en sciences sociales, Paris, France. He was the recipient of the 2018 International Cosmos Prize.

Translator **Anne-Marie Feenberg-Dibon** is Professor of Humanities at Simon Fraser University, Vancouver, Canada. She studied at the Sorbonne (Licence d'Anglais, DES d'Anglais), France, and the University of California at San Diego, California (PhD in comparative literature), USA. Her publications include articles on the novel as well as translations of six philosophical books and numerous articles.

Routledge Explorations in Environmental Studies

www.routledge.com/Routledge-Explorations-in-Environmental-Studies/book-series/REES

A Green History of the Welfare State
Tony Fitzpatrick

The Governance of Green Urban Spaces in the EU
Social Innovation and Civil Society
Judith Schicklinski

NGO Discourses in the Debate on Genetically Modified Crops
Ksenia Gerasimova

Sustainability in the Gulf
Challenges and Opportunities
Edited by Elie Azar and Mohamed Abdelraouf

Environmental Human Rights
A Political Theory Perspective
Edited by Markku Oksanen, Ashley Dodsworth and Selina O'Doherty

African Philosophy and Environmental Conservation
Jonathan O. Chimakonam

Domestic Environmental Labour
An Eco-feminist Perspective on Making Homes Greener
Carol Farbotko

Stranded Assets and the Environment
Risk, Resilience and Opportunity
Edited by Ben Caldecott

Society, Environment and Human Security in the Arctic Barents Region
Edited by Kamrul Hossain and Dorothée Cambou

Environmental Performance Auditing in the Public Sector
Enabling Sustainable Development
Awadhesh Prasad

Poetics of the Earth
Natural History and Human History
Augustin Berque

Poetics of the Earth

Natural History and Human History

Augustin Berque

Translated by Anne-Marie Feenberg-Dibon

Routledge
Taylor & Francis Group
LONDON AND NEW YORK

earthscan
from Routledge

First published in English 2019
by Routledge
2 Park Square, Milton Park, Abingdon, Oxon OX14 4RN

and by Routledge
52 Vanderbilt Avenue, New York, NY 10017

Routledge is an imprint of the Taylor & Francis Group, an informa business

Translated by Anne-Marie Feenberg-Dibon

Originally published in France as:
Poétique de la Terre. Histoire naturelle et histoire humaine, essai de mésologie by Augustin Berque
© Editions Belin/Humensis, 2014

British Library Cataloguing-in-Publication Data
A catalogue record for this book is available from the British Library

Library of Congress Cataloging-in-Publication Data
Names: Berque, Augustin.
Title: Poetics of the earth : natural history and human history / Augustin Berque ; translated by Anne-Marie Feenberg-Dibon.
Other titles: Poâetique de la Terre. English
Description: Abingdon, Oxon ; New York, NY : Routledge, 2019. | Series: Routledge explorations in environmental studies | Includes bibliographical references and index.
Identifiers: LCCN 2018056690 (print) | LCCN 2019003870 (ebook) | ISBN 9780429259388 (eBook) | ISBN 9780367200534 (hbk) | ISBN 9780429259388 (ebk)
Subjects: LCSH: Philosophy of nature. | Nature and civilization.
Classification: LCC BD581 (ebook) | LCC BD581 .B43713 2019 (print) | DDC 113–dc23
LC record available at https://lccn.loc.gov/2018056690

ISBN: 978-0-367-20053-4 (hbk)
ISBN: 978-0-429-25938-8 (ebk)

Typeset in Goudy
by Swales & Willis, Exeter, Devon, UK

MIX
Paper from responsible sources
FSC® C013056

Printed and bound in Great Britain by
TJ International Ltd, Padstow, Cornwall

Contents

Notice

1 In this book, people's names follow the normal East Asian order: the last name precedes the first name. For example: Mao Zedong, not Zedong Mao; Kurosawa Akira, and not Akira Kurosawa: etc.
2 Transcriptions are in the pinyin system for Chinese; for example: Mao Zedong, not Mao Tsetoung, or Mao Tsetung. For Japanese, the Hepburn system is used, where the consonants are more or less pronounced as in English and the vowels more or less as in Italian.
3 Unless otherwise specified, the translations are those of the author.

Preface

Re-naturalizing culture, re-culturating nature through history

"Re-naturalizing culture, re-culturating nature": this Marxian inspired phrase is derived from one of the themes in the *1844 Manuscripts*, "Naturalisierung des Menschen, Humanisierung der Natur" (naturalization of the human, humanization of nature). I have used it before at the beginning of another book;[1] it obviously haunts me. That book, intended to develop the principles of a *mesology*, started with the idea that the relation of human societies to the terrestrial expanse originates and functions in ways which the classic dichotomy between the subjective and the objective cannot grasp. This dichotomy which we call *dualism* is the origin of the couple of modern conceptual oppositions "nature versus culture", and the resulting separation between the social sciences and the natural sciences. On the contrary, mesology aims at understanding what in a concrete milieu unites in a single reality that which dualism separates into two poles.

I suppose that this is somehow the sense which the expression "re-naturalizing culture, re-culturating nature" sought to convey. I cannot be sure because the expression is not mine, nor was I there when it was used for the first time; moreover there are no accounts of the discussion in which it was used. That took place in the family home in Saint-Julien, in the Born county of Les Landes, probably during the 1965 summer vacation. I was not there because I was doing my military service in Baden-Baden, in Baden-Württemberg. As he sometimes did during the vacation, my father, Jacques Berque (1910–1995), had assembled a few friends from the region (or friends from far away passing through), for discussions of an elevated intellectual order in a relaxed and amicable summer setting. That time Henri Lefebvre (1901–1991), who often came on his way to or from Navarrenx, was there as well as Roland Barthes (1915–1980). There were a few others, but my sisters do not remember their names nor the theme of the encounter, since their role was limited to cooking for and serving these gentlemen. However in later conversations with my father I inferred that the theme of the day included the phrase "re-naturalizing culture, re-culturating nature". He uses it also in one of his books *The Second Orient* (1969), but without any reference to the meeting in Saint-Julien, as well as in his *Mémoires des deux rives* (1989) which includes the following passage, unfortunately not dated (p. 178):

> "Re-naturalizing culture, re-culturating nature": I soon shared this phrase with the black militants of the school for revolutionary executives in Wenneba in Ghana, who endorsed it by acclamation. While it is impossible or almost impossible to translate it without some paraphrase, it nevertheless expressed Marx's still romantic inspiration: "Re-naturalizing culture, re-culturating nature". Revitalized by the sap of the African forest, the expression was romantic indeed and even naively so, as it assumed that technology not only consisted in machines but also in patience and ambitions.

Both the geography and experience of the Japanese milieu, with its orientation towards phenomenology and even more to mesology, led me to use the same phrase a generation later. I adopted the term mesology to translate what the philosopher Watsuji Tetsurô (1889–1960) called *fûdogaku* or *fûdoron* in his work *Fûdo*, published in 1935. First encountered in 1969, this book became seminal for me, and I ended up translating it into French.[2] I explained my reasons in *Écoumène* and will therefore not mention them here. However I must correct an error I made for more than a quarter-century that concerns the origin of the term mesology. For a long time I attributed the invention of the term to Louis-Adolphe Bertillon (1821–1883) who gave it a certain luster, as evidenced in the first edition (1866–1876) of Pierre Larousse's *Grand dictionnaire universel du XIXe siècle*, where I discovered it. I had not yet read in *Études d'histoire et de philosophie des sciences concernant les vivants et la vie* (1968) by Georges Canguilhem (1904–1995) the following passage (pp. 71–72):

> In *Système de Politique positive* (1851) Comte names two doctors he calls his disciples, Segond and Robin. They are the two founders of the *Biological Society*. ... The founders of the Society were inspired by positivism. On June 7, 1848 Robin read a memo *On the direction proposed in a meeting of the founding members of the Biological Society to explain the title they chose*. Robin presented the Comtean classification of the sciences, and outlined the tasks of biology in the spirit of Comte's work, the most important one being the study of milieux, for which Robin even invented the term *mesology*.

So let us give Charles Robin (1821–1885) his due in the history of science. He remains better known for having written a dictionary of medicine with Émile Littré which became the standard text after 1873, and for opposing the election of Darwin to the Académie des Sciences in 1872 in the name of positivism. He claimed: "Darwinism is fiction, a poetic accumulation of unverifiable probabilities and seductive explanations without demonstration."[3] Mesology however withered away when ecology later emerged, since it did not have the conceptual or logical resources to deal with such a vast field; today we could say it was torn between medicine, the social sciences and the natural sciences.

I took up the obsolete term[4] again, however, not in its extended sense and especially not from the same perspective. (The first time I referred to it was in *Le Sauvage et l'artifice. Les Japonais devant la nature*, 1986.) Mesology today has taken in the hermeneutical phenomenology introduced by Watsuji for the study of human milieus. It focuses on a central question: how does one go from the universal objective basis of the Earth, or the environment, to what *is specific to a particular subject*, an individual, culture or species?

Watsuji's answer to this question could be summarized as follows: history gives meaning to milieu, and milieu puts flesh and bones on history. In *Écoumène*, I focused primarily on the second part of this proposition, without closely examining the first. Besides, this study exclusively concerned human milieus, the totality of which forms the ecumene, without considering living milieus in general. Their totality not only forms the biosphere, but also *the biosphere in its cosmophany*,[5] that is, as a *geocosmos*, the general world encompassing all the particular worlds that coexist on the Earth. Considered as such, the biosphere remains only the totality of objective *environments*, the sum of ecosystems. However, that does not take into account the phenomenological dimension of living *milieus* which alone makes them into worlds. Between environment and milieu, there is an essential difference, which was also brought to light independently in the 1930s by Uexküll and Watsuji: Uexküll for living milieus, Watsuji for human milieus. The environment is a universal given (an *Umgebung*, as Uexküll calls it), while the milieu, or ambient world (*Umwelt*), is a unique elaboration of this given. It is *an elaboration of meaning* which is proper to each species (of living beings) and proper also to each culture (for the human species). "Meaning" [*sens*] here and throughout this book is fully taken in its triple and jointly accepted sense of *spatial-temporal direction*, *capacity for bodily sensation* and *mental signification*. It goes without saying that such elaboration is more or less complex depending on the species.

The present book has a double objective. On the one hand, it illuminates how history gives meaning to the world around us, in other words how it creates a milieu out of the environmental givens. On the other hand, it aims to show that this process functions according to the same principles, both at the level of the evolution of life, and in history. I want to show that the homology between evolution and history is the key to the question of meaning, although it is an aporia for dualism. It is the object of mesology: the milieu proper to all history and history proper to each milieu.

In this problematic, the question of the subject has a crucial place. The usual approach limits this question to human affairs, while the rest – the evolution of species in particular – is reduced to mechanistic objectification. This bipartite division was established by the dualism of the classic modern western paradigm. From this dualism, which is correlated with and symmetrical to the modern scientific object, emerged the modern individual subject, who became the central actor of our world. They could not exist without one another.

The subject became its own foundation and by cutting itself off from its milieu it opposes a world which then itself becomes an object. This book intends to argue that such a choice was not only not inevitable but also must be transcended. This is our urgent task and needs to be done *Before Nature Dies* (1965), as Jean Dorst (1924–2001) once wrote, and before our own world collapses as a result of abstracting from its *mesological*, and not just its ecological, foundation in nature. Nature has seen worse, much worse!

Part I of this book starts with the question of the subject, to show that abstracting the modern individual subject resulted in a *de-worlding* which will be deadly in the long run, since no being can live without the cosmophany of a common world (*kosmos*). We must *re-world* our existence.

Part II shows that *focusing on the object* which characterizes modernity, ends up stripping things of their meaning, and rendering language unreliable. We have to reposition words and things in the course of their common history, their *growing-together, concrescence*, in other words to *re-concretize*.

Part III shows that to *re-engage* nature and culture, the question of the relation between history and subjectivity has to be addressed at every level of being, from the evolution of the most primitive form of life to the most advanced human consciousness.

Re-worlding, re-concretizing and re-engaging: faced with those three urgencies, western thought today is weighed down by its linguistic origin which yesterday made its strength: the basic subject–verb–complement structure. This structure has oriented our logic (with the subject–predicate model), our metaphysics (with the identity of being), and even our science (with the in-itself of the object), all founded on the double principle of identity and the excluded middle, that is, on the foreclosure of the symbolic. Examples such as the Japanese language, whose basic structure is different, or the tetralemma developed by Indian scholars, which systematically includes the middle, show us the way: overcoming the impossible contradictions of modernity cannot be done without help, *both logical and ontological*, from the great civilizations of Asia.

Our ambition is to give the lie to Kipling's famous saying: "Oh, East is East, and West is West, and never the twain shall meet."

Notes

1 Augustin BERQUE, *Écoumène. Introduction à l'étude des milieux humains*, Paris, Belin, 2000.
2 WATSUJI Tetsurô, *Fûdo. Le milieu humain*, Paris, CNRS, 2011 (1935).

3 As related by Armand de Quatrefages in *Les émules de Darwin*, Paris, Alcand, 1894; cited by Alain PROCHIANTZ, *Qu'est-ce que le vivant?*, Paris, Seuil, 2012, p. 114. Darwin was finally elected in 1878, four years before his death.

4 *Mésologie* no longer appears in the *Petit Larousse*, even though it could be found in the first edition (1906) with the following definition: "Part of biology which deals with relations of milieus and organisms."

5 Cosmophany is defined as the *appearing of world* (cf. A. BERQUE, Cosmophanie ou paysage, pp. 741–744, in Dominique GUILLAUD, Maorie SEYSSET, Annie WALTER (eds.) *Le Voyage inachevé … À Joel Bonnemaison*, Paris, Orstom/Prodig, 1998). Like "ecumene" the term can be understood in a particular sense (dealing with the world that belongs to a certain species, or a certain culture), or in a general sense (the totality of specific cosmophanies, i.e., the lifeworld, *Lebenswelt*, specific to life on Earth).

PART I

Re-worlding

1 The poem's reversal

§ 1. In the woods of Touraine

Touraine is a region in France where supposedly the purest French is spoken. You go for a visit to check it out. At the edge of a forest, a remnant of a thickly forested Gaul, you meet a man who tells you: "I am René Descartes." If you are a native speaker of the same language, you will know that this man has also said, in other circumstances, "I think, therefore I am." Going into the woods, you begin to meditate on this strange homology, that is, to realize that *independent of circumstance*, this man expressed his identity with the immutable personal pronoun "*I*": "*I* think ...", "*I* am." You yourself have done the same thing, and in other circumstances as well. You have never referred to yourself any differently, except when talking to your young children; then instead of saying "I" you said "Papa" or "Mama". Sometimes you may have said "your servant" instead of "I", but that was only for fun.

You arrive at a clearing and as you keep thinking, you realize that "*in some circumstances*", you may not have to be "I" but somebody else. It could even be, as Rimbaud assured us, that "I is another". Then who is he? A being that would not be this *persona prima*, this "first mask" of the actor imposed by the French language? A universal mask for all, the "I", although all are different, all singular, and never speak under the same circumstances?

You suspect here some magic trick, and you take out of your bag the *Discourse on Method* to get to the bottom of it. And there you read "I understood from this that I was a substance whose entire essence or nature consisted solely in thought and that, in order to exist, has no need of a place, nor depends on any material thing."

There it is, you tell yourself: this first person, the "I", exists independent of any site, of any material thing, because it keeps its identity regardless of circumstances, of what the body is like, feminine, masculine, young or old. It is ubiquitous, universal and transcendental. It is indeed a *persona*, an interchangeable mask behind which anyone can hide the singularity of his face, whoever he is ... and so? And so?

You go into the darkening woods and some strange, unnerving feeling engulfs you. This home, which is so familiar to your being, this "I" that welcomes you since your childhood, begins to disorient you and becomes strangely *unheimlich* [note of translator: German for uncanny, eerie]. Is it possible that Descartes, that native of Touraine, might have been fooled when he saw his "substance", his "essence", his "nature" and his unique identity in something that turns out to be just a mask?

§ 2. In the Hercynean Forest

You have been walking for a long time and you still have not reached the other edge of the woods, which turns out to be a vast forest. It is perhaps that Hercynean Forest mentioned by Caesar in *The Gallic Wars* (VI, 26–28), reputed to have covered a territory of more than sixty days' travel, and housing a large diversity of peoples and languages. On the other side of the Rhine, the French "je" becomes "ich"; on the other side of Alps, it becomes "io"; on the other side of the Pyrenees, "yo"; and on the other side of the Channel, "I". However, what all these masks have in common is to remain identical to themselves in every circumstance, and in whatever place.

You can verify this constancy throughout the entire forest by going further and further east. The "ich" was followed by the "ja", then by the "Я", but the principle remained the same. This principle of ego-centeredness, or *Ichheit* as Meister Eckhardt called it in the 13th century, was established in the Indo-European languages originating from a millennia-old root, *ego* (me), which gave the Sanskrit *ahàm*, in Greek *egô*, and in Latin *ego*. But what is this ego-centeredness? For a European it seems self evident: it is a self that can say "I". Perhaps I am proceeding too thoughtlessly? Our immediate English and German neighbors do not have this very French pair, "je" and "moi". The Germans will only talk about *ich* or *das Ich*, while the English translate *moi* by *self*, translated by the French as *soi*. The passage in Descartes quoted above continues as follows: "En sorte que ce moi, c'est-à-dire l'âme, par laquelle je suis ce que je suis ..." ("and so this self, that is the soul, that makes me into what I am ...") becomes for instance in Locke "that consciousness whereby I am myself to myself" (*An Essay concerning Human Understanding*, II, xxvii [1690]).

Let us retranslate "that consciousness whereby I am myself to myself". A troubling metamorphosis! Does that say the same thing? Let us return to Rimbaud who seems to have been right when he claimed in his letter of 15 May 1871 to Paul Demeny that "I is another".

In Barbara Cassin's commentary for the *Vocabulaire européen des philosophies*, Étienne Balibar proposes (pp. 656 ff.) three interpretations of that famous formula. The first refers to God: for instance, the theosopher Franz von Baader's idea that reverses Descartes' *cogito ergo sum* into *cogitor [a Deo], ergo Deus est* (God thinks me, therefore He is); the second has to do with the body or the unconscious – the "I think" becomes the Lacanian "it thinks", and the third refers to the impersonality of language itself, in other words the realm of symbols.

In Part II of this book, we will consider this problem in greater detail. However I would like to mention a common Japanese form which seems to be its direct translation: *omowaremasu*. This form of the verb *omou* is used for example in reports where in French you could say: "one might consider ..." or even simply "I think that ...", but in a more polite mode than the basic form of *omou*. To say *omowaremasu* may indicate the passive voice (thus agreeing with Baader's *cogitor*); or a possibility (one could think so), spontaneity (doesn't the "it thinks" have something spontaneous about it compared with the revisions of consciousness?); or respect, which forbids you from expressing your opinion too personally.

There is a little bit of all these here, but mostly respect – the attempt to be more impersonal – when you finish the presentation of your views with *omowaremasu*, the verb coming at the end of the sentence in Japanese. In order not to put yourself too far forward, out of politeness for your audience, you pretend that it is not you who is doing the thinking; it is as if *it* was thinking out loud: is *it* the Japanese language or the world around you, or intersubjectivity?

Let us continue on this path towards the Japanese language; it seems promising.

§ 3. Among the flat-faced people

In the end, you have walked for much more than sixty days and you arrive in a region where people have flat faces; anyway this is how they appeared to a young Roman, Lucius, a thermal baths engineer. He is a character in Yamazaki Mari's manga, *Thermae romae* (2009), which tells his surprising story. They speak a language called *nihongo* in which "people with flat faces" are called *hiratai kao zoku* (I, p. 22); however this is probably only the translation from Lucius' mother tongue Latin: *natio plani vultus*. They see themselves differently from how he saw them and prefer to call themselves *harakara*, "those with the same matrix", whose nature, essence or substance, as the Cartesian "I" would say, is to speak that language.

In this language that Lucius found so surprising, and that we call Japanese today, *there is no "I"*. "I" is always another, depending on circumstances. Moreover, there are no personal pronouns. If today you encounter such terms in that language, it is because grammarians in the Meiji era, with the help of Chinese characters, pinned European categories on realities that are very different from what those categories designated in the French example. "Pronoun" for instance became *daimeishi*, that is "word" (*shi*) in place of name (*mei*), which is actually the literal meaning of *pronoun* (from the Latin *pronomen* used by Varron in *De lingua latina*, 8, 45). Modern German has a homologous transposition in *Fürwort*, literally "word (*Wort*) for (*für*) [serving as a word]"; however, while this was possible in German, an Indo-European language with pronouns like Latin and French, it was not possible in Japanese, a language from a totally different family. The pronoun was created by being derived from an imported category, translated and superimposed on something that had practically nothing to do with it. Before the Meiji restoration, in the feudal society of the Tokugawa dynasty, there were dozens of ways to talk about the first and second person, according to the circumstances of the dialogue. For the second person there were more than one hundred ways and even today there are around sixty ways, in spite of modern democratization. To speak of "pronouns" in this case obviously makes no sense. Japanese persons do not multiply into dozens of people where we have only six (I, you, he/she, we, you and they.) They simply do not separate their existence from the various circumstances to which they adapt their verbal expressions.

Exactly the same principle is at work – i.e., taking circumstances into account, but far less elaborately – when you say "Mama", or "Papa", to refer to yourself when you talk to your little cherub.

The linguist Suzuki Takao, in his *Words and Culture* (*Kotoba to bunka*, 1973), tries to show the logic of these circumstantial expressions of the person. There is nothing primitive or childish about them. They simply signify that in the Japanese language the focus is on the concrete effective relations between speakers, while in languages with pronouns, it is on the virtual exchange of what the Greeks (and after them the Romans) called the "theatrical mask", *prosôpon* (or *persona*), that is "I", "you", etc. Suzuki challenges the use of the *ninshô daimeshi*, the "personal pronoun" that is only a transfer from European grammars and is irrelevant here. He shows that in reality the Japanese language tends to express the person nominally. For instance, when you talk to your teacher, you will designate him by "master" (*sensei*) and not by "you" (*anata*), which would be very rude. This would also be true the other way around: in front of his class in grade school (and certainly not in other circumstances) the teacher will designate himself by that same term, *sensei*, used by the pupils to talk to him. Indeed, another correlative principle of the Japanese language is to tend towards empathy, to put oneself in the place of another person, especially if the interlocutor is a child. Again, it is what you do in the same situation when you designate yourself as "Mama" or "Papa".

Then you might say, what about this *anata* ("you") you noticed in passing? Isn't this a personal pronoun like the other terms you meet right away in Japanese, *watakushi* ("I"), *kare* ("he"), *kanojo* ("she"), etc.?

It is true that in today's Japanese those terms are roughly equivalent to our personal pronouns, but only roughly. Let us call them pseudo-pronouns, whose use is very different from our personal pronouns. First of all, people usually do not use them in ordinary dialogue; this is not the case in French where the verb is always preceded by the "je" or "tu" (or "vous"). It is the same in English or German, while in Spanish and Latin, the language prefers to play on verb forms to distinguish persons, but still makes distinctions. The important thing is that a word like *watakushi* in its very nature does not correspond to our "I" for the very reason that in English you have only the pronoun "I" to represent yourself (which was already the meaning of the Indo-European *ego*). *Watakushi* only began to mean "I" in the 15th century; until then it had meant something "private" as opposed to the "domain of the sovereign", which meaning it still carries today. The sinograph 私 which represents it in *on* reading (derived phonetically from Chinese) is homophonous to 市 (municipal) i.e., *shi*; therefore we distinguish in speech a *watakushiritsu daigaku* (private university) from the *shiritsu daigaku*, which could mean either "private university" (私立大学), or "municipal university" (市立大学). In writing of course, thanks to sinographs, no confusion is possible between those that belong to the private sphere (*shiritsu* 私立) and those that belong to the municipal domain (*shiritsu* 市立), which in turn is part of the public sphere (*kôritsu* 公立).

Whatever the case may be, this has obviously nothing to do with our notion of "I". Descartes' "I think" was not established "in private", but in front of the entire world.

Nor do the terms representing the second person have anything to do with our "you", singular or plural. *Anata* for example ("you", which is represented by the sinographs 貴方, i.e., "precious orientation") has as its primary meaning "over there". You could say it signifies the great distance between your merit and mine, which is minuscule. The term began to be used in this sense during the Edo period (1603–1867). Apart from the ordinary meaning of "you", which is not used for a respected person, it is most often used by women addressing their husbands, where an English spouse would say "darling" or "honey".

In short, these terms never have the universal currency of our personal pronouns. Just like the very different expressions they most often replace, they always correspond to a particular situation.

In other words, even though it is possible to literally render Descartes' "I think therefore I am", the canonical translation of which is *ware omou, yue ni ware ari*, nobody would say this. Such a sentence can only be used in a particular context: a translation of or commentary on *Discourse on Method*, while in French it is absolutely commonplace, at least in its form. Since it expresses the establishment of the modern subject in its aseity, that is the foundation of the self independent of any circumstance, it becomes more of a linguistic problem. The Japanese language is actually at the opposite extreme of the transcendental universality of the subject designated by "I". The Japanese language rejects this subject in its very functioning. *Ware omou* has nothing in common with the nature, essence or substance of the Cartesian *cogito*. It is exactly the same thing, not coincidentally, that God told Moses (Exodus, 3: 14) with *atte aru mono*, Japanese for "I am who I am", the translation from the Latin in the Vulgate *sum qui sum*, from the Septuagint Greek *egô eimi ho ôn*, and the Hebrew *ehyéh asher ehyéh*).[1] It says nothing we would expect but it would be closer to "what there is because there is" or to "that which, being, is." You have to be a convert to understand such talk! Clearly, what doesn't go over very well in other languages is precisely those self-evident facts! The Japanese language would have preferred the less transcendental *sum ubi sum*, "I am where I am", or even a purely immanent *sum id, ubi sum*, "I am that where I am" (§ 4).

So what does Japanese have in place of the "I" of the *cogito*, which pretends to be my being?

§ 4. At the Lyceum of the Stagirite and what happened there

The same natural beings can evoke very different images in different worlds; in other words, the same subjects can have different predicates as the case may be. Take the wolf for instance, a carnivore from the family of canines (*Canis lupus*). In the Vallée-aux-loups, in his Velléda tower, Chateaubriand wrote his *Mémoires d'outre-tombe*, in which Claude Lévi-Strauss liked to read what he thought were certainly the most beautiful sentences in the French language. The park where Chateaubriand himself planted the trees still attracts those in love with that language; the name of the place, "valley of the wolves", makes them dream. Lukeion, situated north-east of Athens, and a place where wolves (*lukon*) lived, had a different destiny. A gymnasium with the same name was built there later, which in French became the *Lycée*. This is where Aristotle gave the young wolves of that time philosophical training: Theophrastus, Diodorus of Tyre, Strato and many others. Let us not mention the Latin version *lupanar* [note of translator: brothel], except to say that *lupa* (she-wolf) also meant "prostitute". Moreover, Aristotle's school was called *peripatetic*, from *peripatein*, to walk around, which he liked to do lecturing in his public evening courses.

Certain images are fleeting while others become part of history and sometimes even govern it, giving rise to what Gilbert Durand called the "semantic pool" (*Introduction à la mythodologie*, 1996). The teachings of the Lyceum were at the origin of this pool, because it determined the relation of modern European civilization to the world, leading to the state of the planet today. The essential motif of this semantic pool is the notion of the subject. It is therefore worth returning to it through different avenues.

Let us start with a recollection from the *lukeion* (*lycée* in French), the famous words in the *Aeneid* (II, 49) that Virgil places in the mouth of the priest Laocoon:

> *Equo non credite, Teucri!*
> *Quicquid id est, timeo Danaos et dona ferentes*
> Beware of the horse, Trojans!
> No matter what, I fear the Greeks
> Even when bearing gifts.

Timeo is the first person singular of the present of the verb *timere* to fear. This first person cannot be mistaken for any other, neither singular nor plural, since the verb varies with each of them: *timeo, times, timet, timemus, timetis, timent*. Here the verbal form *timeo* necessarily and exclusively expresses an "I" (the *ego* implied by this form). The same thing happens in the other tenses (the imperfect, future ...) and in the subjunctive mode. The first person, the speaker here, is Laocoon, the subject of the verb to fear. Without any doubt, we know he is the one who is scared because we immediately recognize the basic structure of our grammar: SVC, subject / verb / complement, I (S) / fear (V) / the Greeks (C). We also immediately recognize the basic structure of our logic, which links the subject S with the predicate P; here: I (S) / fear the Greeks (P).

Let us now listen to little Hanako (a favorite character in Japanese grammars), who went to the zoo with her papa:

– *Kowai!:*	(I am afraid!)
– *Nani ga kowai?:*	(What are you afraid of?)
– *Tora ga kowai:*	(I am afraid of the tiger)

Tora is the tiger and the particle *ga* tells us it is in the position of subject. The subject of which verb? There isn't any in this sentence. *Kowai* is a qualifying word related to Hanako; it qualifies the emotion she feels: fear. This is strange from the point of view of French grammar and even logic. Hanako says she is afraid but the subject of the sentence, *Tora ga kowai*, is the tiger. Indeed, we do not have the basic structure SVC. The subject feeling scared is Hanako, not the tiger. Do we at least have the logical structure S/P (S as P)?[2] That remains to be seen. The predicate is what we say about the subject. However, here the predicate indicating a feeling of fear (*kowai*) does not concern the tiger (*tora*), but Hanako. The tiger is just the motif of her fear. Moreover, all she said was: "I am afraid" (*kowai*), without us yet knowing why. If this is really a predicate it is truly odd. Its very structure cannot do without Hanako, the speaker, even as the sentence says nothing about her, as is also the case of *Kowai!* and of *Tora ga kowai* as well. The real subject of what is felt here (fear) is the existing Hanako, and not the substantive *tora*.

Let us try another path. What if we translated *kowai* by "horrible, frightening" as for instance the *Nouveau petit dictionnaire japonais-français*, edited by Hakasuisha (1994), invites us to do? Leaving aside the fact that a tiger cannot be "frightening" by itself (it can only be frightening for a subject abstracted or hidden by our language), wouldn't that just leave *tora ga* (S) / *kowai* (P): the tiger (is) frightening?

Certainly, this is what it would look like. However, a Japanese–French dictionary only aims at making it possible for the Japanese to express themselves in French, or for the French to manage in Japanese. It does not say what it means in Japanese for the Japanese. In the first utterance of

Hanako's, *Kowai!* (I am afraid!), the tiger is only implied by the girl who is afraid; but her father who does not know this (at least not in the grammatical example) is obliged to ask her what the problem is. In this utterance *kowai* is not really the predicate of *tora*, but of Hanako instead.

This is uncomfortable for an English person who remembers that what we normally understand as logic, since Aristotle,[3] is based on the identity of the subject. What is the subject here: the tiger or Hanako? Or both? A difficult question indeed, which gets worse as we continue with Japanese grammar. We learn that a predicate as simple as "Hanako is sad", or *tora ga* (S) *kowai* (P) can only be said by Hanako herself, designating herself by her first name as little girls or young women in love often do. Others have to say *Hanako wa kanashisô da*, "Hanako looks sad". You can say "I am sad" (*kanashii*, or pedantically, *watakushi wa kanashii desu*) but you cannot say "she is sad" (*kanojo wa*) *kanashii* (*desu*); you must say "she seems sad". Now you have arrived at the heart of the problem; in Japanese you are obliged to link the utterance to the concrete existence of the speaker, you in this case. Words cannot stand alone; language cannot turn on itself indefinitely. As in Gödel's famous *Unvollständigkeit* theorem, it needs a referent external to itself to remain consistent. The referent is you; concretely it is you who makes a statement about Hanako; however since you are not she, you cannot say she is sad. You can only say she seems sad.

In other words, in Japanese the basic structure S/P, as such, that is in the binary relation of S and P, does not work.[4] It needs to have a third referent that is not in the utterance itself but is outside language, i.e., the existant that speaks in certain circumstances unspecified by the utterance.

Let us elaborate with a more exciting example, a haiku by Oshi:

風鈴の	*Fûrin no*	The wind chime
ちひさき音の	*chiisaki oto no*	A delicate sound
下にゐる	*shita ni iru*	I lie beneath

In the *New Almanac of Haïku* (*Saishin haiku saijiki*, 1977, vol. II, p. 149) of Yamamoto Kenkichi, this poem is classified with "seasonal words" (*kigo* 季語) of summer. As we will explain more fully in § 28, haikus must include a seasonal word, which here is *fûrin*, wind chime. In the summer it hangs on a tree branch in the garden, or on a projecting beam, its clapper equipped with a small strip of paper to create flutter, chiming at the least puff of wind; by synesthesia it is refreshing when it is stuffy inside the house. Unlike what the translation says – in which I respect the order of the original verses – the Haiku sentence is grammatically quite ordinary and tells us literally: "wind chime – of // small sound – of // be below"; which means: "to find oneself beneath the little sound of the wind chime". The form of the verb *iru* (to be somewhere) designates no one. It can be interchangeably I, you, she/he, we, you or they. In Japanese there is no verb inflection according to person, nor does it have, as we have seen in § 3, a personal pronoun to confirm the person as in French. In any case there is neither one nor the other, neither inflection nor pronoun. The morphology does not indicate whom it concerns. What is said explicitly is the chiming of the bell, the "finding oneself (*iru*) of some latent presence" underneath it, or feeling the cooling wind as if it were underneath, that is, some being which would be *what it is where* it is: under the delicate sound of the wind chime ... in other words pure immanence: *Sum id, ubi sum* (§ 3)!

The English language is forced to specify who the subject of the verb *iru* is, and thus to translate *shita ni iru* plausibly by *I* am beneath, *I* find *myself* beneath. As we have seen above, the ordinary functioning of the Japanese language implies the existence of the speaker, so we can be pretty sure that we are dealing with the person who is reciting the poem. Indeed it is I, the speaking subject, who is beneath the bell. Unless we engage in some bizarre contortion, the poem then would begin in English with the subject of the verb *iru*, that is, "I": "I lie beneath the delicate sound of the wind chime."

However, the English language invents the subject in question to make it explicit and so formally excludes any other virtual subject, listener or reader for instance. In other words, English makes subjectity[5] absolute by restricting it to a single actor – a virtual one at that – in the poem. It modifies its meaning, since other actors are reduced to the status of objects, attributes or predicates, in short all that surrounds my existence. It is quite different on the one hand to start with the environment (the sound of the wind chime) and end evoking a latent existence (the possible subject of the verb *iru*) and on the other hand to start with the existence of the subject in question, the "I", and then outline the attributes of that same subject (that is, he finds himself beneath the wind chime). This is a profound restructuration of reality, which in this scene as well as in the word order of the poem, is literally turned upside down.

It is impossible for the English language not to effect this *reversal of the poem*, because it is obliged to provide not only a grammatical subject for the verb *iru* but also to give a logical subject to the predicate "finding oneself beneath the delicate sound of the wind chime". This does not happen in the Japanese language which can easily do without the two basic structures SVC and S/P.

– So then what organizes that language? What is its basic structure and what is its logic?

– This will not be easy to uncover because ever since Meiji, grammarians and logicians, both Japanese and foreign, have enmeshed it in garments from the West. Fortunately, not all of them.

Notes

1 Clearly, the Septuagint version is not literal, because it means "I am the being". Nevertheless, it could have been literal, because the transposition from Hebrew into Greek, and into Latin and French, does not pose a structural problem; however, in Japanese, this transposition is structurally impossible.

2 Throughout this book, including in the formula (((S/P)/P')/P'')/P''' ... (§ 8), I will use the sign / not in the sense used by a mathematician (for whom S/P would mean S divided by P), but specifically in the mesological sense of "as" (*als* in German, or in Japanese *toshite no* [*P toshite no* S]).

3 To be sure, he was not speaking of logic, but of analytics and dialectics. The Peripatetic Alexander of Aphrodisias, during the reign of Septimius Severus (193–211 CE), was the one who unified both under the term *logikê* (implying *technē*, the art of reasoning).

4 KUMAKURA Chiyuki, *Nihonjin no hyôgenryoku to kosei. Atarashii watakushi no hakken* (The capacity for expression and individuality among the Japanese. The discovery of a new I), Tokyo, Chûôkôron, 1990, bluntly says about this (p. 19): "Japanese does not have the subject–predicate structure of European languages", a structure which according to him made possible the omniscient narrator of the modern western novel (which the nouveau roman has challenged).

5 Note of the translator: as Professor Berque explains in Chapter 5, he prefers to use "subjectity" rather than "subjectivity" which has connotations that need to be excluded.

2 Destinies of the subject

§ 5. Why do we say "subject"?

Is there a more treacherous word than "subject"? No, because in several ways it may mean both the thing and its contrary: the *subject* of the logician (that which is in question) is the object of the physicist (that which the science studies, and makes its *subject*); for the grammarian however, the *subject* of the verb is not at all the same as its direct or indirect object. For the logician, the *subject* is opposed to the predicate (what is said about the subject), but for the philosopher, the subject who is the physicist is the opposite of the object he observes. The philosophers' *subject* is his own master and subjects the world to his gaze, but the *subjects* of Louis XIV were subjected to their sovereign's gaze ...

This is a cascading series of reversals that affects the derivatives of the word "subject": subjectivity for instance, which is often confused with *subjectity* – the fact of being a subject, to have a self (*selfhood*, as Locke § 2 could have said); but subjectity or selfhood, in the sense of capacity of proprioception, or even self mastery, can also mean its opposite, because to be subjective means to lack objectivity, or being subjected to passion – to be *subject* to whims or lovesickness for instance. The human subject is par excellence endowed with subjectity but, equally par excellence, devoured by passion. *Subjection*, whether to passion or tyranny, puts a final touch to the confusion about the *subject* which definitely eludes those who want to grasp it; this is why philosophy has pondered the question of the *subject* for a long time, at least in our circles.

Why then stick to such a misleading word? Simply because the history of our language makes it necessary. There is no other reason: it is both necessary, since this history offers no alternatives, and satisfactory as we have used it since the Late Empire. The English word *subject* is derived from the Latin *subjectum*, the translation of the Greek *hupokeimenon*. *Subjectum*, used in the late sense of subject by Minneius Felix Martianus Capella, the 5th century CE African writer, is the neutral past participle of *subjicere*, to put below, to subject. It is the "recumbent", that which lies under being as its foundation. *Subjectum* was a literal transposition of the Greek *hupokeimenon*, neutral past participle of the verb *hupokeimai*, to lie under, be the base of something, which was substantivated as *to hupokeimenon*. Since ancient Greek had neither the equivalent of the English "subject", nor that of "object", *hupokeimenon* did not include the idea of subjectivity, but only the sense of something lying underneath and serving as support. This corresponds more or less to the *shita ni iru* of the haiku we saw in § 4: the *hupokeimenon* lying underneath is subject to or supports something else. In Aristotle's *Physics*, it is the matter (*hulê*) of the thing, the substratum of accidents in the changes it experiences, and in the *Categories*, it is its essence (*ousia*) which supports the specifying predicates. The notion of lying beneath something, *hupo tini keimai*, is close to that of *substance*, a word derived from the Latin *substantia*, from *substare* (to stand beneath) translated from the Greek *hupostasis* (the fact of lying beneath), which in turn became *hypostasis*, the synonym of *substantialization*. In the European philosophical tradition the metaphysical relation between substance and accident is analogous or even homologous to the relation between subject and predicate.

Without dwelling any longer on this peculiar history, let us go straight to the point that concerns us.

§ 6. To emerge or to sink?

Let us compare two utterances, banal in their form if not in their ontological principle which is to affirm one's own existence in the world: "Je m'appelle René Descartes" ("I am called René Descartes"), and "Nishida Kitarô to môshimasu" ("I am called Nishida Kitarô"). The first utterance in French right away posits the identity of the speaker as "je". Then he projects the identity onto itself ("je" calls for "me", which is "je") and then gives his official identity: the name in the following order: 1. first name (of the individual person) 2. last name (of the family to which this person belongs). [Note of translator: this comment does not pertain to the English language, so the French construction has to be used here.] On the other hand, the second utterance starts by reversing the order: first the last name and second the first name and then following the connecting particle *to*, it ends with the verb expressing the speaker's action: *môsu* (the polite form of "to say", *iu*), in its even more polite form, required in this exchange of civilities inherent in the introduction of oneself to strangers. Additionally, this verbal form as we have seen in § 4 is impersonal: there is no trace of the "I" in the entire utterance.

It is immediately clear that the respective priorities of those two existences are not the same. They are inverse. Moreover, even reversed, the two utterances do not say the same thing: the "I" and its reflexive play, are absent in the second one. None of this is neutral, at least if we agree that language is not extraneous to being; however, this point will have to be justified.

– Justified, but to whom, in the name of what?

– To the "I", which in today's world claims universality as if any other than the Anglophone mode of life should be foreclosed, or relegated to folklore. In this book I hope to show the importance of keeping this claim within the bounds of a certain world that best serves only those who belong in it. For all others, this world is certainly not the best. We may even have to change our world for the sake of a deeper concern than grammar. Let us start by illuminating a few of the characteristics of this "I" to see if they are not actually singular rather than universal, and historical rather than eternal.

The being who can utter "I am called René Descartes", as well as "I think therefore I am", at least has no doubt about his "I". In a way, he expresses and focuses his subjectity. What then can we say about the being who says: "Nishida Kitarô to môshimasu"? The least one can say is that he does not focus his subjectity on himself. Since we are dealing with a being who is no less human, and therefore no less endowed with subjectity, his self can only be the opposite of "focused", in other words, diffused in the circumstances, or more generally the milieu in which the utterance is spoken. This being is an *ambient* rather than a subject opposed to the object.[1]

Here we have in a few words the symbol of two diametrically opposed philosophies. At least at the verbal level, Descartes' utterance seems to belong to an *egologue* (the "I" narrative) and Nishida's to a *mesologue* (a milieu narrative). We have seen in § 1 the passage in the *Discourse* where Descartes posits his subjectity as independent of any milieu. What in Nishida's opposite extreme would correspond to this? In a passage of *Basho* (1926) where he expresses what is at the heart of his philosophy, there is a term, which in this case I would translate by milieu:[2] it refers to the absorption of the subject (identified with being) by the predicate (identified with nothingness). The sentence in which he states this idea most explicitly presents the remarkable feature that the two verbs have no subjects. To give all due respect to the French language, in the following translation I had to struggle to provide them with subjects by adding words in brackets:

> [One] might think that here the predicate becomes the foundation. As [the latter] swallows in the predicate plane, without changes, the subject plane which until then was being, [the predicate] ends up also including the meaning of the will that subsumes the general in the particular.[3]

What Nishida expresses here in philosophical language corresponds more or less to the *shita ni iru* of the haiku we saw in § 4: the "finding oneself beneath", a *hupokeimai* invested by the "predicate plane" (what is said about the hypothetical "subject" of the verb *iru*). In Nishida's terms, "that which until then was being" (of the subject in question) is swallowed up in the predicates of certain circumstances (finding-oneself-beneath the sound of the wind chime). In other words the subjectity of the subject becomes immanent to the scene. The elision and implication of the subject go together. The same momentum allows Nishida to withhold subjects from the sentence above: even grammatically, subjectity is a given and immanent since the preceding sentence already refers to it: as the foundation (*kitai* 基体), in other words the *hupokeimenon*.

In one case we have the transcendence of the subject (the "I" of the *Discourse*), in the other the immanence of the subject (Nishida's swallowing up): this is more or less shown in the two preceding examples. However, since the second one is rather obscure, let us take a simpler and more general one. When you say in Japanese "samui" it may mean either "it is cold" or "I am cold", or generally both at the same time, inextricably. *Samui* is a word indicating a quality which is sufficient unto itself, without verb or subject. It functions as a predicate expressing an ambience, which implies a subject feeling this ambience but without verbalizing it. We have here a common expression in the Japanese language which in Nishida takes the form of a philosophical thesis:[4] the subject is swallowed up in the ambience or the milieu (*basho*). On the contrary, in English the expression of that same ordinary scene evokes the Cartesian dualism, where the "I" of "I am cold" cannot be confused with the "it" of "it is cold". "I" is the subject par excellence, who speaks in the first person, while the "it" is impersonal, devoid of any subjectivity, if not of grammatical subjectity. As Benveniste has shown[5] it is indeed a "non-person" because only the pronouns of the first and second person designate interlocutors. Only "it" can predicate things; "I" and "you" cannot do this because they are deictic, shifters that propel discourse into language. This is the problem in languages that have pronouns. But it is also inherent in languages with transcendence if I may use this expression. In practice, the transcendence of the "I", the subject par excellence outside of any milieu, has as its exact counterpart the capacity of the language to make the "it", and therefore the milieu, into a pure object, apparently abstracted from any subjectity. When you say "it is cold", it is an utterance as formally objective as is "it is minus thirty". Of course this is a fiction because you actually do exist and feel the cold but it does not engage your existence. You are just as isolated from it as is the cabin of an airliner isolated from the −60 degrees of the outside temperature. In Japanese, the problem is the reverse: the speaker's immanence in the language makes it difficult to abstract from his subjectity as well as from the situation in which he finds himself. *Samui* is used in informal speech; on a case by case basis, other situations may require a greater degree of politeness, starting with *samui desu*, or even silence. In other words, discourse surfaces in language, which has led some people to profess that a strictly objective argument is impossible in Japanese.[6] However, the fact that physics, the paragon of objective science, can be expressed in Japanese just as well as in French needs some explanation.

Let us clarify our point of view before continuing. I adhere, up to a point, to the Sapir–Whorf hypothesis according to which language influences thought, but I certainly do not believe that thought is determined by language. Of course there is influence, but the relation between the two terms is complex, contingent and historical. It is more about *going-with, concrescence, co-creation and motivation* than about causality. For instance, Sanskrit is a language with pronouns where, as we have seen in § 4, the "I" exists (*ahám*), but where also an ontology emerged that could be considered the opposite of modern Europe's. However, in modern Europe, doesn't Ortega y Gasset say something in Castilian that seems to go beyond modernity, to be the opposite of the abstract *cogito: yo soy yo y mi circunstancia?*[7] We will return to this in Part II. For the moment we will continue with the emergence of the modern "I", whose being gradually became detached from its milieu. Descartes' statement quoted in § 1 illustrates this emergence. Of course this did not happen overnight. It has distant roots while its effects have not yet fully surfaced. The *Discourse on Method* is not the *cause* of its emergence but one of its paradigmatic expressions.

Among those distant premises Aristotle's thought plays an important role on at least two points. One pertains to logic, the origin of our double notion of subject and predicate. In the *Categories*, the Stagirite posits that true and false exist only where there is a combination (*sumploké*)[8] of a name and a verb, in other words of a subject and a predicate. He also posits that predication stops at the primary essence (*ousia prôtê*), which takes on the figure of the individual, and which is not part of any "category", in other words it cannot be the subject of any predication as such. To quote his translators:[9]

> The primary substance is the end point of predication, the maximum of sensory determination. The maximum of sensory determination is singularity. To complement the distinction between primary substance and secondary substance, Aristotle establishes a hierarchy founded on the hegemony of the primary substance in reality and in determination, which signifies the hegemony of the singular over the general.

This individual without predicate will thus be the pivot of any judgment on the world while he himself would by definition be exempt from it. That is not all.

Correlatively, in his *Physics*, book IV, Aristotle produces a definition of place (*topos*) which severs any identitarian link between the thing (the logical subject in question) and its place because the thing is mobile while the place is not. There is only the "immobile receptacle" the *angeion ametakinêton*, where occasionally the thing can or cannot be found, but which every time demarcates it exactly. Hence the definition of the place as *to tou periechontos peras akinêton prôton*, "the immediate immobile limit [of the thing]" (IV, 212a20).

In other words, the thing is A and place is non-A, since they are distinct and separable. This *topos* is radically different from the *chôra* Plato discusses in the *Timaeus*, the milieu whose indefinable identity is both the imprint and the matrix of relative being (*genesis*), and so participates in it ontologically.[10] An individual cut off from any milieu and strictly delimited by his corporeal envelope: those are indeed the essential characteristics of what I call "the modern ontological *topos*", or MOT.[11] It concerns the subject as well as the dualistically conceived object (that is the logical subject, § 5). Of course, the MOT, the figure of contemporary individualism and dualism, does not correspond in every point to the Cartesian "I", which as we have seen in § 2 stands for a mind and not a body; however, modern materialism quickly forgot this distinction. *Habeas corpus* (§ 12)! You shall have the body, in its *topos*! Isn't it through the body that the MOT subject is defined?(§ 32). Isn't he its owner?

Of course, the roots of the MOT do not only go back to Aristotle but are multiple and composite and include Christianity as well. For instance the MOT owes to St. Paul the assurance that the absolute (God) will reside in each of us,[12] and to St. Augustine (*Confessions*, X, 24) the conviction that He is really there: *manes in memoria mea* [*Domine*], *et illic te invenio* (you remain in my conscience [Lord], and this is where I find you). With such a tenant, the *cogito* was surely justified in thinking that it had been set free from any terrestrial milieu!

But this *u-topia* (the negation of place) has still more profound roots. According to Paul Jorion,[13] the birth of the subject owes a great deal to the "Pythagorean power grab" whose distant consequence in the 18th century was to mistake mathematical modeling for reality.

This interpretation is supported by comparing it to the purely geometric image of perspective, *costruzione legittima* of Alberti. Panofsky has shown that the "legitimate construction" symbolizes the emergence of the modern subject,[14] reduced to a theoretical point outside the *image*. Outside the image is so to say symbolically outside the milieu. It is the inverse of what constructions in Chinese painting were aiming for: to pull the subject *inside the landscape*,[15] that is into a milieu. That is the same operation whose principle is outlined in Nishida's philosophy, just as, conversely, it is the principle of "legitimate construction" in Descartes' philosophy.

§ 7. Without link or milieu: the subject outside the world

The theme of the exteriority of the modern subject with respect to the world, and vice-versa, has given a particular meaning to existence, a word derived from the Latin *existere*, to stand outside. It is homologous to the Greek *ekstasis*, the station outside oneself (which became ecstasy in English). We know that Heidegger made much use of these terms to signify that being-there, *Dasein*, is outside the self (*außer sich*), among things (*bei den Dingen*). In other words, it goes beyond the envelope around the MOT. In that sense, existence is *ek-sistence* outside the MOT, towards and in the world (*in der Welt*). However, the emergence of the modern subject has inversely resulted in *ek-sistence* outside the world, retreating in the MOT. Heidegger rebelled against this and in particular against Cartesian dualism which made the world into an object outside our existence, while for Heidegger it is our *ek-sistence* itself.

Clearly, with respect to the subject, existence can have two opposite meanings: either it is detaching oneself from the milieu to assert itself in the MOT, or going beyond the MOT to become milieu. In mesology as I understand it, the question is not the choice between one or the other interpretation: both are complementary since those two aspects, the *ek-sisting-ouside-the-world* and the *ek-sisting-towards-the-world* are historically combined in a double movement, of *trajection*, the to-and-fro we will often mention in what follows.[16] My thesis is that modern dualism has stopped this to-and-fro; it is "the focus on the object" (§ 14), the *ek-sistence* that individualizes the MOT in detaching it from the world. We will start our analysis with the latter aspect and show the multiple facets of the constitution of the MOT.

In principle, the MOT produces a being that abolishes all foundational limits except the one defining himself, the *peras prôton* that coincides with the contours of his body, and possibly with those of the objects he owns[17] (this is one of the sources of capitalism, § 12). Of course, previous milieus knew all sorts of limits, and these were often sacred. However, today psychoanalysis can say the following about the MOT, without actually naming it:

> What can we say about the soft nihilism that flourishes in our democratic societies where the all-powerful ideology of well-being triumphs unconditionally? It has led to the foreclosure of religion, the repression of sexual difference and the denial of violence, led to frenzied individualism and immoderate demands for all sorts of rights. Does this mean that for humans the foundational limits of the human condition are becoming unacceptable?[18]

Clearly, this quote brings together the idea of democracy and loss of limits. Ever since its emergence, democracy has pushed for the abolition of limits which put people in different categories and ends up circumscribing each in the unique mold of his/her own MOT. This will have been the history of freedom: the *ek-sistence* of the subject outside the limits imposed by tradition. From his own inner self, the subject now bends the exterior world to his gaze, as the "legitimate construction" has given him the right to do. Up to the vanishing point, far away and even into the infinite, the MOT subject orders everything from his own individual perspective, without any intermediary limit, without allowing for any obstacles, for wouldn't that be illegitimate?

In the exterior world, now measured by the geometry of Cartesian coordinates (the x, y and z axes), space is purified into a simple extension (*extensio*), and things become objects, each allotted its proper place in a legitimate construction. This kind of place, fictitiously described as objective, although Panofsky has shown that it is entirely derived from the construction of the modern subject, is called *Stelle* by Heidegger. It is only a site, a random spot, a purely circumstantial *topos* devoid of any ontological link with the thing, and thus with the *ek-sistence* of *Dasein* towards the world. In *Being and Time* (1927), Heidegger finds there the *Platz*, the place belonging to the thing in its existential relation with the subject. Later he will distinguish it from the *Ort*, the place which is no longer a random place in Descartes' "pure space" (*reiner Raum*) or Newton's universal space, but instead a site where "spatiation" (*Räumung*) occurs, the emanation of a singular space coming from the thing which goes beyond its *Stelle* (its *topos*) to welcome *Dasein*.[19]

However the MOT subject does not recognize the spatiation of things, because this would engage his being in the *chôra* of an *ek-sistence* that would be the inverse of his own. He is not "close to things" but sees them from a distance like an aviator form above.[20]

Remember that in Plato, the theme of the *chôra* – the milieu – is both ontological and cosmological. It partakes of the *kosmos* – the general world order. In the onto-cosmology of the *Timaeus* relative being (*genesis*) could not be separated from the *chôra* without at the same time being de-worlded, in other words, without losing its being-in-the-world (its *in-der-Welt-sein*, in Heideggerian terms.) The de-worlding is the equivalent of what Heidegger calls *Entweltlichung*, the loss of world of the modern subject. Soon after the emergence of the modern scientific paradigm, this is probably the destiny sensed by Pascal in his famous saying: "The eternal silence of those infinite spaces frighten me." The MOT is the de-worlding, de-globalizing and ultimately inhuman division inherent in the dualism and the mechanicism of this paradigm.

Three centuries after Pascal, Hannah Arendt more explicitly denounced "the flight from the world for the self" characteristic of modernity. Fleeing the world is however not a new theme in human history. Since antiquity, all great civilizations have known it and in particular the first centuries of Christianity, and also at the same period but with an entirely different view of nature, the Three Kingdoms and Six Dynasties in China as well. In Greek this movement was called *anachôrêsis*, the "return to the *chôra*", that is the countryside outside the city walls. In Roman antiquity, the world is literally equated with the city whose foundation was accompanied by the digging of a sacred hole, the *mundus* that according to tradition was an Etruscan ritual. It enabled the world of the dead (the *infernum*, "below") to communicate with the world of the living. Furthermore it involved the drawdown of a chunk of the sky, the *templum*,[21] which became the cosmic axis whose pivot was the city. Besides, the Latin *mundus* was the equivalent of the Greek *kosmos*, which also means the world, ornament and ordering. To reject the city meant rejecting the world, which was what anchorites did when they retreated to the desert (in Greek *erêmos*, hence the English *hermit*). While Syria or the Thebaid had deserts in the present sense of arid zone, in premodern Europe, the "desert" was just a deserted, that is, uninhabited region, a wilderness. The etymology of the Latin *desertum* means "detached" (from *serere, sertum*, to weave, to attach). To go to the desert was indeed *to detach the links that make a world.*

Eremitism is one of the distant origins of the modern individual, in as far as it brought human beings face to face with nature. However, even though they rejected the world in terms of worldly relations, the Fathers of the Desert as well as the retiring mandarins of the Six Dynasties remained *woven* in a cosmicity profoundly permeated by a higher meaning, desired by God for some and the expression of Dao for others.[22] On the other hand, the MOT subject, in his desert is quite detached from any higher meaning. Like the Cartesian "I", the modern subject founds himself according to his own consciousness.

§ 8. The self foundation of the self, or of the world?

Many cosmologies, as for instance the Bible's Genesis, or the Australian Aborigines' story of the *Tjukurrpa*, create the world by giving things names. According to the *Laozi*, the name-bearer (*youming*) is the mother of the ten thousand beings.[23] Human speech produces a cosmos; it establishes a world, *the* world to which each of us belongs. This is how Christianity thought of the world, expressing it through symbols: *In principio erat Verbum*, "in the beginning was the Word" (the Gospel according to St. John, 1: 1), and "God said: 'let there be light, and there was Light' [...] God called the light 'day' and the dark 'night'" (Genesis, 1: 3–5) etc. This may be so, but what is happening when we say "I am called René Descartes"? Or when we say it the other way around, "Nishida Kitarô to môshimasu"?

As we saw earlier (§ 5), in the first case the identity of the "I" is reflected onto itself. That is the meaning of the reflexivity of the human subject, its "being for oneself". So what? Taking hold of one's self is not necessarily a function of the MOT alone. As we will see in Part III, it is characteristic of each living being to different ontological degrees. To be sure, language allows for a degree of reflexivity infinitely superior to that of the simple grasp of the self (proprioception) of living beings. Even so, self mastery could be understood in a very different sense from that implied by the "I". This is the case in Nishida, for whom the *jikaku* (the self awakening) is effected by *basho*, also called the world (*sekai*), which unlike the substantial self (*ga* 我), he equates with absolute nothingness (*zettai mu*). Why is the world associated with nothingness? Because it is predicative, not subjectal or substantial. Incidentally, let us remember that for Aristotle the predicate is also insubstantial. However, for Nishida the world is downright nothingness but also predicative: it is "the predicate world" *jutsugo sekai* 述語世界. Once more, this vision is diametrically opposed to Descartes' and could be called its enantiomer. Nevertheless, we are still dealing with the question of the identity of the "self" (*ji* 自).

I should emphasize that in mesology if not in grammar, the self in question is first pre-verbal or sub-verbal. To be able to say "I", in other words to predicate the self as "I", presupposes the concrete existence of the self as a substratum or the subject of predication through language which calls it "I". The latter again posits the "I" at a different level as the subject of a subsequent verbal predication, as in "am called René Descartes".

What does the *môshimasu* of the Japanese language reveal about this "selfhood" (ipseity – the identity of the subject with himself)? Clearly, just as the French language goes together with Descartes' ontology and vice-versa, so the Japanese language matches Nishida's philosophy of nothingness. Going together here is almost literally the case: the verbal predicate "calling" (*môsu, môshimasu*) lacks any subjectity (there is no "I" here), yet it says and creates all by itself the identity – being of the self – of the human subject "Nishida Kitarô". This looks like the very thesis of Nishida's philosophy of nothingness: the predicate, first a relative nothingness (*sôtai mu* 相体無) in as far as it negates being (*u* 有, i.e. substantial being), from subsumption to subsumption of the subject in the predicate, that is from negation to negation of what subsists of being, ends in absolute nothingness which negates itself. From the self-negation of nothingness emerges its contrary: being. In other words this means that from the nothingness of the predicate *môshimasu* emerges the substantiality of the subject "Nishida Kitarô".

Of course there is both a logical and mesological flaw. Logically, the negation of being by relative nothingness can only be the infinite regression of being; it can never turn into a negation of nothingness by itself. We will see in Part II that this logical flaw is a mystical leap which consists in making P absolute. This mysticism originating in Buddhism is of the same order that inspired St. John to write that the Word (in other words the predicate: what is said about a subject) is God, in other words, the absolute subject, the *ipso facto* absolute word. Mesologically it amounts to forgetting that a predicate does not come out of nowhere: it is uttered by a subject in the flesh, and here he is indeed called Nishida Kitarô.

Similarly, although inversely, there is also both a logical and mesological flaw in the self-grounding of the *cogito*. Logically it is the same thing as making its subjectity absolute but it does so by what negates it as such: the predicativity of "I" with respect to its existential substratum which is its real *hupokeimenon*. With respect to the foundation of modern dualism, making S absolute also makes the logical subject of science, the object (S), absolute. This is no less mystical than Nishida making P absolute. In as far as it concerns the object, this is just a take-over by scientism: pretending to hold the truth of objectivity (the pure object, that is, absolute S). Scientism forgets that science, the predicate given about the object, presupposes the existence of the subject itself (the scientist) who proffers the predicate while leaving aside his own being in the name of objectivity. The mesological flaw is self-evident: the *cogito* forgets that in as far as the subject is alive he has a body whose soul (or consciousness as Locke says) is inseparable during his life, and whose life is necessary for him to pronounce predicates such as "I am" about himself.

For mesology here also, the question is not about opting for one or the other of these opposite conceptions of the roles of S and P, Descartes' or Nishida's. They are complementary because each seizes only one aspect of the trajection that in reality grounds the being of the human subject, that is *the subject predicating itself*.[24]

The phrase "I am called" in the English language reveals the difference with the *môshimasu* of the Japanese language: the subject ("I") predicates the self ("me") as itself ("I"). This is a double movement, a trajection: on one hand *the assumption of the self, the subject as "I", predicate* (I assume my being as the "I") on the other hand, going backward, *the hypostasis of the "I" in identity with oneself*, the substantialized identity, i.e. the subject, that gives me the foundation on which to say "I" about myself as predicate. Similarly in Husserlian phenomenology the noesis produces noema which founds the noesis in return and so on; or in Marx representation and reification are imbricated. In this case, "I" can count on "me", banking on it to proclaim its own identity. There is nothing absurd about this because it is not about an identity fixed at the moment t_0, but it concerns a process which like any reality unfolds in time, in the historiality of a time that is its own. The phrase "that is its own" signifies that this phenomenon does not unfold in a universal and abstract time, but in a time concretely linked to the phenomenon in question and interacting with it. This is about history and milieu, in other words, trajectivity, and not about Newtonian absolute time and space.[25]

If we represent the subject by S and the predicate by P, we could formulate the trajective reality of human identity as follows: (((S/P)/P′)/P″)/P‴ ... and so on. We can see that each new instantiation of the predicate P (P′, P″, P‴, etc.), in other words each awareness or utterance by "I", is founded on the earlier reality S/P (S taken as P), and is indefinitely replaced in the position of subject or the substance S, S′ [i.e. S/P], S″ [i.e. (S/P)/P′], S‴ [i.e. ((S/P)/P′)/P″], etc., pending predication by P, P′, P″, P‴, and so on.

To say that the S/P reality is indefinitely replaced "in the position of the S subject" is to say that P (or P′, P″, P‴, etc.) is indefinitely hypostasized (substantialized) in S′, S″, S‴, etc. Trajection is an indefinite sequence of assumptions and hypostases of these assumptions (S/P, S′/P′, S″/P″, etc.), of noeses and noemas, of representations and reifications, that constitutes the reality of the human being, the subject predicating itself.

At the theological level of the absolute, this *self-as-other*, this otherness as predicate of the self that constitues the reflexivity of the human subject, is reflected in St. John's utterance: "In the beginning was the Word", in other words P ... "and the Word was God", in other words S. This is not different from human reflexivity[26] deifying, or at least fetishizing its own ontological structure. In short, it is the key to the theophany of the Subject which predicates itself as an ouroboros, as proclaimed on Mount Horeb, *Ehyéh asher ehyéh* (§ 3) ... Since mesology does not deal with the absolute, it has no answer to this hypothesis and cannot even exclude the inverse metaphor, which in the Bible says that God created man in his image ...

Short of the absolute, we will see that in human milieus, the same principle of trajection is valid for what we consider reality. In other words, the reality of things is *trajective.*

§ 9. Mastery of the self, or awakening of the self?

We have seen the schematic opposition between the two language/thought couples: on the one hand the English language based on the substantial Cartesian subject of the "I think, therefore I am" and on the other hand the lack of substantiality of the "predicative world" of Nishida, based on going-together rather than causality. This opposition could easily look like a coarse illustration or even a caricature of the Sapir–Whorf hypothesis. The following facts will show that the reality hiding behind those examples is much more complex.

Nishida's thought owes much to Buddhism and to Zen in particular. The theme of nothingness (*mu* 無) in his work is but another name for the Buddhist emptiness *kû* 空 in Japanese, *śūnya* and *śūnyatā* in Sanskrit. The theme of emptiness has been most developed in the Buddhism of the Mahāyāna, the Great Vehicle. This thought is diametrically opposed to the substantialism inherited from the Greeks and Christianity yet it was written in Sanskrit, a language that is no less Indo-European than French or Greek. As such this language has pronouns, including the "I", *ahám*, it inflects the verb according to the person, and it has a comparable subject/predicate structure. This has not stopped Indian grammarians, and in particular the greatest of them Pānini (560–480 BCE), considered the legislator of Sanskrit, from establishing a grammar that is not based on the subject/predicate relation, and in which the sentence does not function by focusing on the subject. How is this possible? Because, to quote the Sanskrit scholar Michel Angot,[27] "grammar is just the language's consciousness", it is not language itself. One could say that if language is the subject S dealt with in grammar, then the latter is its P predicate. If so, the S/P (S as P) is born historically from a certain reality that can influence thought in ways that are not immanent to the language, but contingent, and steers it in a given direction. This seems to be the case in the development of Mahāyāna, first in India in Sanskrit and from there in China and in Chinese, to end up in Japan and Nishida's language. In the course of this development, certain tendencies were reinforced in translation since the structure of Chinese (a Sino-Tibetan language) has nothing in common with Sanskrit, nor with the structure of Japanese (an Altaic agglutinative language), which has no similarity with the structure of Chinese. At the end of this long history, Nishida's thought, while owing a great deal to Zen, the current stemming from Mahāyāna, shows multiple analogies with the characteristics of the Japanese language.

From what we have seen so far in French, it clearly appears to favor the hypostasis of the "I" into "me". Inversely, Japanese constitutes a fertile ground for the Buddhist notion of *anātman*, the absence of *ātman*, or the non-existence of a mind that could be considered as the substance of the subject. *Anātman* in Japanese is *muga* 無我, "non-me". This word sums up an intellectual current which requires abandoning the notion of ego. This will seem difficult in English, but is self-evident in a language where "I am" becomes *môshimasu*, without "I" or "me" and without a verb inflection indicating the first person. Since the most important European languages all are similar to English in that respect, they probably have facilitated the hypostasis of the notion of subjectity and the concentration on the human subject alone. The reverse is true for Japanese. In Part III we will see the possible consequences in a domain to which *a priori* it has no relation: the theory of evolution. Let us just mention quickly that even a thinker like Heidegger, who wanted to destroy what I have called here the MOT, and even the naturalist Jakob von Uexküll, who greatly influenced the Heideggerian conception of worldliness (*Weltlichkeit*), remained dependent on the European tradition in this respect. On the one hand Japanese naturalists such as Imanishi Kinji did not hesitate to recognize animal subjectity, at the opposite extreme of the Cartesian view of the animal machine; however on the other hand Uexküll and Heidegger considered the animal as "poor in world" (*weltarm* so *ipso facto* poor in subjectity),[28] while the human, essentially the only real subject, is "world-forming", *weltbildend*, according to Heidegger.

I have no doubt that the attribution of subjectity to the human alone owes much to the historical alliance of the Indo-European languages with Christianity, derived from the biblical tradition, which was conceived in two Semitic languages, Hebrew and Aramaic. This was not the case in India.

If the acquisition of the self by the subject seems practically tautological in English, what is the case in Japanese? What Nishida understands as *jikaku* 自覚, (self awakening) points to the predicate, and not to the subject as we have seen. The latter can only become self-conscious in as far as s/he is in the world, not *distant* from the world like the MOT aviator, but immersed in and engulfed by it (*botsunyû* 没入). Undoubtedly, this will evoke Zen and notions such as *satori* (enlightenment). Since precisely nothing can be said about that, it will not clarify anything. Our problem is to understand how a human being can be conscious of himself, as all humans are barring accidents, both when the history of the culture has emphasized *ek-sistence towards the world* and when conversely, as with the MOT, it has emphasized *ek-sistence* outside the world.

Trajection dismisses both tendencies back-to-back: in reality they are complementary. In any case, the respective paradigms, that is, the highest forms of the culture, are very recognizable. We are familiar with the complex one that produced the MOT, the classic modern western paradigm, where the Copernican and then Galilean shift are combined with the experimental method (Bacon), Cartesian dualism and mechanicism (Descartes), the absolute space of Newtonian cosmology, etc., in short everything that converted the world into an objectal universe, abstracted from the existence of the subject.[29] On the contrary, the dominant tendency of Japanese culture has been to found the world in that very existence. In that case, things endowed with feeling will be what concretizes the establishment of the self. This ideal is literally expressed in a very ancient principle of Japanese poetry, already explicitly posed by the Manyôshû: to give things the task of saying what one feels (*mono ni yosete omoi wo nobu*). Human feeling investing things has been the guiding thread of Japanese aesthetics during its entire history (§ 23). It originated concepts such as *mono no aware*, the emotional stirring of things, where the boundary is lost between what one feels and what things themselves would feel. The subjective and objective merge in the reality of a unique feeling.

Not only did the Japanese language favor such openness to the phenomenal but it also contributed to it by its own evolution. Around the same time that in the West the MOT was established, we see the extraordinary development of impressives, onomatopoeias that imitate not only sounds (called *giongo* 擬音語 or *giseigo* 擬声語) but all sorts of states[30] (called in this case *gitaigo* 擬態語); for example, sparkling stars in a calm summer night becomes *chira chira*; the stroll of an elegant woman in her kimono becomes *shanari shanari*, etc. Mangas systematically use this device, and graphics add to the effect. While impressives are indeed words, they function entirely without syntax. They are the phenomenon as given, that which the syntax immediately decomposes to recompose in another verbal order. In short, Japanese impressives are the embodiment of what verbalization has to forgo: seizing the phenomenal nature of existence itself in its unity, which language disintegrates through its syntax. A famous aphorism by the Indian monk Nāgārjuna (2nd–3rd century CE) denounces it as *gantā na gacchati*, "the walker does not walk".[31] Indeed, by distinguishing the actor of movement (the walker) from the movement itself (to walk), and as we would say, by distinguishing the subject from the predicate, language falls into an aporia. Nāgārjuna could not have denounced this in *shanari shanari*; here the agent of the movement, a rather snooty pretty woman, is the movement itself (a gracious walk), expressed in just one single word! Impressives overcome the very aporia of the verbal, i.e. rendering an atmosphere in a linear fashion; and this is why in the Japanese language they are the most difficult for a francophone to master, because French tends to verbalize as much as possible.

There is no need to add that impressives elude the distinction between subject and predicate. They embody the awakening of a sentiment that assimilates them and which is something different from the grasping of the self by the *cogito*. It is not by accident that at the same time impressives were developed, what was to become the haiku emerged, the poetic form for *the creation of a certain atmosphere* rather than the articulation in words of a scene centered on a subject. Just as we speak of dew point in meteorology,[32] the haiku stands in a sense at *the language point* that precedes the total verbalization of the world by the subject. Its brevity but especially the obligatory inclusion of "seasonal words" (*kigo*), insert the haiku into the world, even beyond words. We will see this in greater detail in Part II.

Here too, mesology refuses the alternative between on the one hand, self mastery by consciousness ordered verbally, and on the other hand the self awakening that gives the impression of a larger and more profound state than consciousness. Human reality is *both* this and that. Consciousness predicates our impressions in linguistic terms, but as Lakoff and Johnson[33] have shown, even our most elaborate concepts only make sense because our flesh interprets or in other words predicates them in actual carnal terms: impressions. In circular fashion, that is to say trajectively, we have here the transfer of our flesh into our consciousness and the hypostasis of consciousness into our flesh. Beyond that, it even means the transfer of the Earth into human subjectity and the hypostasis of consciousness into our Earth. This at least is what my book will try to show, without falling into mysticism.

§ 10. The imposition of the subject in Japan

Modern Japanese has avoided the ambiguities we saw in § 4 because it translates "subject" by a series of distinct terms. The psychological or subjective subject of the *cogito*, opposed to the object, is *shukan* 主観; the logical subject, opposed to the predicate, is *shudai* 主題; the grammatical subject is *shugo* 主語; the subject as his own master is *shutai* 主体; the subject subjected to a sovereign is *shinka* 臣下, etc. All these words are composed of two sinographs. They associate *shu* 主, which means "master, principal" in the determinant position and in the position of the determined, another sinograph: *kan* 観 (look), *dai* 題(question), *go* 語 (word) *tai* 体 (body), etc. The above order shows "principal look" and "principal question", etc.

Nevertheless, those terms were not imposed overnight during the Meiji restoration. Even during Nishida's time (1870–1945), they were not as firmly implanted as they are today. The notion of the subject implies, to say the least, the very foundations – the *hupokeimena* – of the modern western worldview and cannot be imported like a manufactured product. To translate the term that crystallizes it, specialists needed to engage in much difficult thought; moreover a political strategy as well as the restructuration of the language had to take place, the likes of which are no longer imaginable. I will describe this briefly.

Yanabu Akira, who became a translation specialist, crowned his work with a book entitled *Modern Japanese Thought*,[34] mostly dedicated to the question of the term *shugo* and its relation to the particle *wa* (*ha*),[35] which marks the subject. Stunning for the layman, Yanabu's thesis is that "the 'subject' has been fabricated by translation";[36] and not just any translation: it is primarily that of the text of the Meiji Constitution, promulgated in 1889.

– Now look here! This text, the "Imperial constitution of the empire of Japan" (*Dai Nippon teikoku kenpô*), was an act of national and even imperial sovereignty par excellence!

– Certainly, but it was inspired by western, and in particular Prussian, models not only for many of its themes, but also for the concepts that embody them, and even the structuring of sentences that express them. As Yanabu has shown, the result was some extraordinary prose that was different from written or spoken everyday language, and practically gibberish. The main editor Itô Hirobumi, had closely followed the preliminary German draft due to the Prussian jurist, Karl Rössler. For instance, the first article of the preliminary draft declared: *Das Kaisertum Japan ist eine auf ewig unteilbare Erbmonarchie* (the Japanese empire is a forever indivisible hereditary monarchy).[37] The German starts with the subject (*the Japanese monarchy*); the same thing in Japanese *Nippon Teikoku*, followed by the particle *wa* that marks the subject. However ordinary Japanese does not often designate the subject. To place it up front by marking it with *wa* (S+*wa*) was revolutionary for the time. Today we no longer realize how revolutionary it was, since such structuring of the written language is now commonplace, the model having been instituted by the Constitution itself and then disseminated through the compulsory teaching of the "national language", *kokugo* 国語[38] and translations of western literature. In legal language especially it has become normal to start sentences with S+*wa*.

Of course, it was already necessary before Meiji to translate western texts, beginning with the first contacts with Portuguese and Christian proselytism, thus initially from Portuguese and Latin. The Latin nominative generally became *wa*. The Japanese grammar by João Rodriguez (1561–1634) notes that the particle "signals the will to indicate, present, underline and enumerate the terms to which it is attached".[39] However it does not indicate that the S+*wa* needed to be normative. It only started to appear in "Dutch studies" (*rangaku*) that under the Tokugawa were intended to spread knowledge of European sciences and technologies. Towards the end of the shogunate a thinker like Nishi Amane (1829–1897), who studied in the Netherlands and to whom Japanese owes, among others, the word *tetsugaku* 哲学 (philosophy), systematically renders the subject–predicate structure from Dutch as S+*wa* … in Japanese.

Nevertheless, before Meiji, such prose remained unusual, emphasizes Yanabu. What made it more commonplace were the school textbooks in compulsory education which systematically used the S+*wa* … structure. For Yanabu, "this is where consciousness emerged of what will later be called *shugo* (subject)".[40] Mozume Takami who initiated the movement to unify the written and spoken (*genbun itchi*) language, begins his manifesto of 1886 as follows: *Bunshô wa* … (The sentence …). *Wa* here is written in katakana (ハ and not は)[41] to emphasize its importance. However, apart from school textbooks and the law, it took a long time for the S+*wa* structure to prevail, especially in the press. Much of the written language remained in Chinese (*kanbun*), read in the order of the Japanese language, very different from Chinese, usually without the *wa* mark since in Chinese it does not exist as such. Moreover, even when a word clearly indicates the subject in Japanese, it is not necessarily marked by *wa*. This is for example the case with the canonical translation of "I think, therefore I am": *ware omou, yue ni ware ari*, which is written as 我思う、故に我あり.

By the way, influenced by Japanese, where *ari* can be spelled 在り or 有り, the usual Chinese translation[42] of that same sentence is *wo si, gu wo zai* 我思、故我在. Since *zai* is a locative, for any ordinary reader this implies that the *cogito* is a *Dasein*, which is incoherent as further down the text says that it needs no place to exist. To say this in Castilian, *zai* can only be *estar*, to be somewhere, and not *ser*, to be. And so do ideas travel around …

§ 11. The torments of the predicate subject

Let us return to Japan. Taught in the schools, the imposition in the "national language" of the S+*wa* structure, in other words, of the subject–predicate structure, exacerbated an old problem, the distinction between *wa* and *ga*, the particle that also marks the subject but can also coexist with *wa* in the same sentence. Here is the grammarian Mikami Akira's (1903–1971) famous example: *zô wa hana ga nagai*, the elephant has a long nose, literally "elephant-*wa* nose-*ga* to be long". In this sentence there appear to be two subjects. Although a very different language,[43] the exact same structure can also be found in Chinese, as in the example quoted earlier in § 8, *neige ren zui dadade*, "this man has a large mouth". In both sentences "nose" or "mouth" are respectively the predicates of the subject "elephant" or "that man" as well as subjects of "to be long" and "to be large". While in Chinese grammarians have no qualms about speaking of a "sentence with a predicate subject", *zhuwei weiyu ju,* this is not the case in Japanese where the problem has still not been solved. The reason it is not solved is the particles *wa* and *ga*, which do not exist in Chinese; their role needs to be specified as it is unclear.

In school Japanese, you can say *Watakushi wa Dekaruto desu* (I am Descartes),[44] but you can also say *Watakushi ga Dekaruto desu.* It does not mean the same thing, and the intonation is also different; in the second case the translation has to be "It is I who is Descartes". In the first case *watakushi wa* ("I") is unstressed and is even omitted in everyday language, which would simply say *Dekaruto desu,* in other words only the predicate "to be Descartes". However, in the second case it cannot be omitted where the strongly stressed *ga* insists on the singular identity of the subject *watakushi*: it is I and nobody else who is Descartes; correlatively *Dekaruto desu* is unstressed. Between the sentence with *wa* and the one with *ga*, the subject and the predicate are practically reversed, to the point that by reversing the order of the first sentence into *Dekaruto wa, watakushi desu* (Descartes, that is me) you could obtain the exact semantic equivalent of the second one.

In other words we need to put into perspective our familiar notions of the subject and the predicate. Japanese is one of the languages where simply applying them could be inadequate. Mikami believed that the notion of the subject (*shugo*) was useless and recommended that the couple subject–predicate (*shu–jutsu*) be replaced by that of theme–predicate (*dai–jutsu*). In the above example this would be: 1. theme-*wa* (elephant), 2. predicate (nose-*ga* to be long). However, the problem is far from solved[45] because in this sentence, "to be long" still has a subject "nose".

Let us leave grammar behind to ask the question at a more general and more political level. We have seen earlier that "subject" in the sense of master of oneself became *shutai* in Japanese. According to Kobayashi Toshiaki[46] the term was used for the first time in 1887 by the scholar of Buddhism Inoue Enryô (1858–1919), in his *Principles of Philosophy* (*Tetsugaku yôryô*). Understandably, it was difficult to get used to it in a culture where the term that is the most current rendition of our "I" *watakushi*, had meant "vassal" (i.e., the subject of a sovereign) until the Muromachi period (14th–16th centuries). Even today it continues to mean "private" (as opposed to "public" *ooyake*); indeed until the last war the civic ideal was "to abolish *watakushi* and to dedicate oneself to *ooyake*" (*messhi hôkô*).[47] In the twenties of the Meiji period,[48] the term *shutai* enters the law, introduced by German jurists with the term *Subject* in the sense of the "subject of rights". In philosophy it is only in the first decade of Shôwa[49] that *shutai* will tend to replace *shukan*, for instance in Nishida, under Marx's influence, to avoid confusing subjectity (*shutaisei*) with subjectivity (*shukansei*). In Watsuji, starting with the first page of *Fûdo* (1935), human subjectity (*ningen no shutaisei*) is posited as the founding condition for the distinction the author establishes between *fûdo* (the human milieu) and *shizen kankyô*, the natural environment as the object of science.[50] We will consider this distinction more thoroughly in Parts II and III. In his major work, *Ethics* (*Rinrigaku*, 1937–1949), which he bases on hermeneutical phenomenology, Watsuji pays no less attention to Marx than to Heidegger. One of Nishida's disciples, Miki Kiyoshi, initiated the transition from *shukan* to *shutai*. In a series of articles starting in 1927, he published a critique of German idealism associated with *shukan*, while exalting Marxian materialism that he associated with *shutai*.

While subjectity is generally associated with individuality (*kosei*), in Nishida for example, the Kyoto School professed the immersion of the subject in the predicate, the latter itself becoming the subject. This was the meaning of Nishida's "logic of the predicate"[51] that reversed the Aristotelian logic of the subject's identity. In the context of that period, the theme drifted towards ultra-nationalism. Nishida himself went so far as to assimilate the predicate, i.e., universal and absolute nothingness, the source of being, to the imperial regime.[52] Another disciple of Nishida, Tanabe Hajime, slightly changed that perspective in his *Logic of the Species* (*Shu no ronri*, a collection of articles published after the war.) He made the predicate, in other words society, nation or the state, into an obligatory intermediary between the singular (the individual) and the universal. He ended up investing the species with subjectity which, given the militarism at that time, easily led once again to the affirmation of *Kokutai* (the national body) over *shutai*. As a result, the Kyoto School was discredited after 1945 and its members relieved of their university functions. Nishida himself had already died, less than two weeks before Hiroshima. Miki, decidedly too much of a Marxist, was thrown in jail in March of 1945 and left there to languish until his death, a month after the surrender.

After the *Kokutai* years, the postwar years were marked by the "controversy over subjectity" (*shutaisei ronsô*). Could historical materialism still acknowledge this subjectity recovered under American occupation? Marx and Freud had to coexist, and the concept of alienation had to be brought up to date ... What then did it mean to be master of oneself? This nagging question kept coming up, and for twenty years, subjectity crystallized all the torments of the time. On March 6, 1965, a student of the University of Yokohama, Oku Kôhei, an activist in the Chûkakuha current, committed suicide at 21. His journal and love letters were published in October under the title, *On the Tomb of Youth* (*Seishun no bohyô*). The book was a great success. Oku discussed the *Manuscripts of 1844*, and spoke of the rift that separated him from the woman he loved, because she militated at the University of Waseda in favor of the Kakamaruha current (the enemy of Chûkakuha) ... the book shows Oku's obsession with subjectity (*shutaisei*):[53]

> I had been using the word *shutaisei* since June 1960.[54] But I think that I needed three years to really understand it. ... There is no doubt that

an overwhelming number of students try to leave aside their own subjectal critique (*shutaiteki hihan*) as student activists – while at the same time ridding their minds of the problem of their subjectal engagement (*shutaiteki na kakawari*) in the student movement, while they take advantage of the July 2 question[55] to conveniently avoid their own subjectal questioning (*shutaiteki na mondai settei*).

§ 12. The MOT hegemony

May shutai *rest in peace* ... Hereafter High Growth with its industrial pollution (*kôgai*) brought the Japanese other worries, not the least of which were the effects of the politics of Thatcher-like neoliberal deregulation (or rather Reagan-like since this was Japan: same thing) which in the 1980s the Nakasone government put in place. As we know Mrs. Thatcher professed that *there is no such thing as society. There are men and women, and there are families.* Interesting for a country where until 1945 the *Kokutai* had been exalted and which believed in the *messhi hôkô* (§ 11) ... It is true that before Meiji the concept of society did not exist in Japan. The term *shakai* to translate the English "society" was first introduced by the journalist and critic Fukuchi Gen.ichiro in the January 14, 1875 edition of the *Tôkyô Nichinichi shinbun*. Before that people used *seken* 世間 or *ukiyo* 浮世,[56] in other words "the world" or "people". Ferdinand Tönnies would have called that a far more medial (belonging to a certain milieu) than universalist perspective, concerning *Gemeinschaft* (community) rather than *Gesellschaft* (society). However, he published his famous *Gemeinschaft und Gesellschaft* only in 1887. For his purpose, Fukuchi unearthed a Chinese term in an anthology of neo-Confucianism, the *Jin si lu* (Reflections on that which is nearby, 1176).[57] There he found the term *shehui* 社会 (pronounced *shakai* in Japanese), whose initial meaning was "the carnival of the gods of the soil". Later the Chinese language re-imported the term in its present sense.

Only individuals exist, and not societies – except those based on a contract between individuals – such is the credo of methodological individualism, the postulate par excellence of the modern ontological *topos*, the MOT. This ideology, arising out of medieval nominalism and particularly influential in Anglo-Saxon thought, is distantly related to Aristotle who did not believe that Platonic ideas had a real existence but considered them simple predicates. It is also related to the Cynic Antisthenes who, mocking Plato, said: "while I actually see horses go by, I do not see horse-ness". An old problem, obviously! And indeed, in Japan too, neoliberalism gets along well with the MOT. They reinforce each other.

The most ancient source of this doctrine is undoubtedly the atomism of Leucippus and Democritus that culminates in contemporary social sciences as mechanicism. There is a homology between the isolated movements of particles considered by classical mechanicism and the "isolated individuals" criticized by Durkheim in his controversy with "utilitarians" like Spencer:

> They presuppose that in the beginning there were isolated and independent individuals, who later only relate to each other to cooperate; for they have no other reason to move across the empty space that separates them and to join each other. However, this widespread theory postulates a real ex nihilo creation.
>
> Indeed it consists in deducing the society from the individual. However, nothing that we know allows us to believe in the possibility of such a spontaneous generation.[58]

Personnel management practices in neoliberalism can be traced back to this homology. *Telerama*[59] for instance could say the following about Jean-Robert Viallat's documentary, *The Killing of Labor*, that dealt with the repeated suicides of managers at Renault or France Telecom: "[He shows] how the logic of profitability crushes social and human relations in the company and kills work itself by reducing it to its sole economic dimension."

Personnel management methods explicitly aim at isolating people, to make each of them a wolf for the other. As the good grandson of Occam's nominalism and grandfather of Mrs. Thatcher Hobbes would certainly not have disavowed[60] this strategy. In short, the idea is to reduce the human to an MOT. Society then becomes just a network of contracts between individuals who, as children of the *cogito*, are ontologically divorced from any milieu.

This contractual basis is obviously linked partially to the reduction of the MOT subject to "the one-dimensional man" denounced by Marcuse. In the System born of mechanicism where the particle is reduced to one function, the MOT subject is essentially deprived of world – abstracted from any *Umwelt*, and so becomes a simple element in the *Umgebung*, the objective environment. At least this is the way it looks according to the System.

– What is this System?

– The System systematizes: everyone according to his/her MOT, which matches other MOTs and all fit together in the System. This is basically what Frank Fischbach denounces when he writes that this evolution produces "both subjects deprived of world and a world reduced to an external reality spread out in space";[61] in other words, reduced to the Cartesian *extensio* of the *Umgebung*. In his critique, Fischbach refers not only to Heidegger but to Marx in particular, for whom capitalism reduces the worker to labor power, separated from the means of production. In fact, according to Fischbach:

> It is precisely this separation that transforms the human being who works and labors into a *subject* separated from the world as a simple abstract depository of pure "labor power" (*Arbeitsvermögen*); this

> separation also explains the movement of real abstraction that goes from the multiplicity of concrete works – to which Heidegger confines himself – to labor power, the social mean making all labor and its products comparable to each other, quantifiable with respect to one another.[62]

In the System the MOT subject has become quantifiable. It is a statistical product resulting from the foreclosure of its milieu, which is now reduced to the *extensio* of the *Umgebung*. There is an obvious coherence between methodological individualism and capitalism. In this economic or ontological system costs are externalized (thrown out into the outer darkness, that is the foreclosed milieu) in order to better internalize profits (that is, to attribute them to an MOT, here the company owned by the happy few). So for example, with the enclosures, the commons that were the *chôra* of small farmers became the *topos* of big property. With just-in-time production, the milieu (the collectivity and the environment) supports the costs which the big companies do not pay.[63] "Just in time": the rest is of no interest to the MOT.

With the alliance between the MOT and capitalism, the historical movement of modernity appears like a huge abstracting of being from its milieu: a *discretization*, (like the decomposition of concreteness). In other words, modernity fetishizes subjects as well as objects, for it imputes a value to them which actually derives from their concrete interrelation. We will say more later (§ 13) about the fetishization of the object. Let us turn first towards that of the subject. In establishing itself, what exactly does the *cogito* do? It virtually usurps the place of the divine to become its own fetish – fetish in as far as it cuts itself off from any kind of relation which in reality allows it to exist. This is the same movement whereby it fetishizes things into pure objects, detached from its existence. We will see what the costs are in § 13.

Reduced to a quantifiable *topos* in the System, what happens to the subject? First the body. Its new status started slowly in medieval England and then blossomed in 1679 with the *habeas corpus* act. It continued with fingerprints and today has led to the biometric passport, described by the jurist Alain Supiot as "detaching identity from any territorial reference". This of course is characteristic of the MOT derived from the *cogito*. He adds: "Biological identification is finally about to supplant civil status in border control organization."[64]

What were identity papers and borders? They were historical instantiations of a mesological fact: human beings are not reducible to the limits of their individual animal body. To be sure it pertains to biometrics; however history supplements the animal body just as necessarily with a *medial body* that is eco-techno-symbolical, that is, its milieu. Identity papers and borders are part of that complex and they go beyond biometrics which does not recognize the spatiation of being – to use the Heideggerian idiom, *die Räumung des Wesens* – because by definition, biometrics is limited to the *topos* of the animal body.

Animal body *and* medial body; in Part II we will return in greater detail to this ontological structure as such. Let us first consider here what being has become after it was foreclosed when modern dualism gave birth to the MOT. To foreclose is not to abolish; it puts outside (*fora* as they would say in Corsica,[65] from the Latin *foris*) in the external darkness, and closes (*claudere*) the door of consciousness. This foreclosure is the lock-out of half of our being, that is, our milieu which is converted into an objectal environment.

This is precisely what the Heideggerian *Dasein* contested. However what could it do against the hegemony of the MOT and the authority of its first premise: I AM MY BODY? This hegemony touches all aspects of our existence, especially since it seems at first to embody our highest values such as respect for the person and his/her freedom. Take for instance the feminist slogan "My body is mine" proposed during the years leading to the legalization of contraception and abortion, and liberating women from the natural and social condition that had made them in this respect unequal to men. Reminiscent of this are pedagogical songs in the school *My body is mine* (bis) / *It is all mine / It is my body, my own / My body is mine / It is quite nice* to which sometimes is added *Mine, only mine*. The intention is to prevent children from becoming victims of sexual abuse. The intention is commendable as are the Neuwirth and Veil laws.[66] Unconsciously however, these expressions participate in the construction of the MOT, which tends to abstract from all natural and social constraints by foreclosing our medial body. For instance, it forecloses the fact that the freedom obtained with the pill presupposes a powerful pharmaceutical industry and the whole system of which it partakes. Ultimately, it inserts the individual consumer in the System that dispenses the pill, sometimes at the risk of one's life as we saw recently with the Diane 35 pill [note of translator: a contraceptive pill].

Such foreclosure makes the MOT subject an inauthentic being. The MOT is inauthentic: not only is his medial body foreclosed but only in as far as the latter develops can he become individualized in his own MOT. He believes himself freer and less constrained by nature and society, the more fully deployed are the symbolic and technical and hence social systems which alone make this freedom possible. In short, the more he is individualistic, the more he is social. However, he does not want to see this. Shut it out!

The foreclosing of the medial body is such a dangerous fantasy! The MOT subject not only forgets he is social, he also overlooks the ecological dimension of his vital milieu. The MOT subject is the creature who most appreciates "nature" when the System shoves more horsepower under his butt, up to 550 for instance in Porsche's SUV Cayenne which takes him to Timbuktu or the supermarket. To liberate oneself from spatial constraints it is better to go by air to Noumea thanks to the 50,000 horsepower and to the 320,000 liters of kerosene of the A380 jet liner. The return flight of the MOT between Paris and New York farts between 2.5 and 3 tons of carbonic anhydride per passenger, which is 40 to 60 times his/her average weight, and more than the 2 tons of CO_2 an average Frenchman emits for his travels in one year.

Hello planet!

For sure, the MOT is just a paradigm for modernity. Like all paradigms it is a predicate (P) and not a reality (S/P). It is up to us to come up with a different paradigm, which is what this book is all about.

Notes

1 I argued this position in Le japonais, pp. 240–250 in Jean-François MATTÉI (ed.) *Encyclopédie philosophique universelle, IV: Le Discours philosophique*, Paris, PUF, 1998. Later the translation of this "ambient" became problematic, until I realized when writing an article for the Japanese journal *Kan* (54, July 2013, Fûdosei kara kangaeru Nihon no shisô [Mediance and Japanese thought]) that I could simply use *tsûtai* 通体, on the model of *tsûtaisei* 通態性, which I had thirty years earlier chosen to translate trajectivity; as a trajectant instead of the French for recumbant (§ 5). The verb 通, reads in Japanese *kayou*, which means "to come and go". The movement of *tsûtai* can be constrasted to the immobility of our subject (*shutai* 主体), a notion which Japan had a very hard time adopting (see further in § 10 "The imposition of the subject in Japan").

2 *Basho* 場所 is usually translated by "place" but in the case of Nishida's philosophy, "field" or "milieu" would be better. The best translation could be *chôra* χώρα, in the Platonic sense, however without the notion of the absolute being, of which Nishida's absolute nothingness seems to be the enantiomer.

3 Or in the original text: 是に於いて述語的なるものが基体となると考えることができる。これまで有であった主語面をそのままに述語面に没入するが故に、特殊なるものの中に一般なるものを包摂するといふ意思の意味を含んで来るのである。 P. 261 in vol. IV of the 1966 edition of Nishida's complete works. (*Nishida Kitarô zenshû*), Tokyo, Iwanami.

4 As noted by NAKAGAWA Hisayasu, Dekaruto to Nishida – futatsu no tetsugaku no gengoteki zentei (Descartes and Nishida – the linguistic presuppositions of two philosophies), *Shisô*, 902, August 1999, pp. 5–19, p. 10. I commented on this text from a slightly different perspective in *Vivre l'espace au Japon*, Paris, PUF, 1982, p. 40.

5 Émile BENVENISTE, La nature des pronoms, *Problèmes de linguistique générale*, Paris, Gallimard, vol. I, 1966, pp. 228–236.

6 The problem is discussed by KUMAKURA, *op. cit.* in § 4.

7 "I am me and my situation"; or analytically "I am me and my milieu in a certain situation". José Ortega y Gasset (1883–1955) proclaimed this famous thesis in *Meditations of Don Quixote* (*Meditaciones del Qijote*, Madrid, Publicaciones de la residencia de estudiantes, 1914, online, p. 43), from a clearly mesological perspective. On the same page he writes:

> Contemporary biology studies the living organism as a unity composed of the body and its particular milieu; so that the vital processes do not only consist in the adaptation of the body to its milieu but also of the milieu to the body.

We will return to this theme in Chapter 8.

8 ARISTOTE, *Catégories. Présentation, traduction et commentaire par Frédérique Ildefonse et Jean Lallot*, Paris, Seuil, 2002, p. 63 (2a5–10 for the Greek text).

9 *Ibid.*, p. 42

10 For a more detailed discussion of the difference between *topos* and *chôra*, see my *Écoumène. Introduction à l'étude des milieux humains*, Paris, Belin, 2000, chap. I, Lieu.

11 Note of translator: here Professor Berque comments on the etymology of the acronym TOM, the French for MOT, which refers back to the Greek verb *temnein*, to cut, cf. atom, anatomy, etc. The English translation of the acronym does not include this Greek root.

12 Corinthians, 15: 28: "And when all things shall be subdued unto him, then shall the Son also himself be subject unto him that put all things under him, that God may be all in all (*ut sit Deus omnia in omnibus*)." Although a language with pronouns, Sanskrit did not produce the MOT, because it did not have to translate the Bible with God's *sum qui sum* to Moses, nor affirm Christianity.

13 Paul JORION, *Comment la vérité et la réalité furent inventées*, Paris, Gallimard, 2009, pp. 262 ff. Jorion's point of view comes close to what we will see from another perspective in § 31, "The liberation of the logos".

14 Erwin PANOFSKY, English version translated by Christopher S. Wood, MIT Press, 1991.

15 On these questions, see my *Les Raisons du paysage. De la Chine antique aux environnements de synthèse*, Paris, Hazan, 1995.

16 The term trajection comes from *trans* (beyond, through) and *jacere* (to throw). It is literally the idea of "throwing oneself" beyond identity, and especially to cross the limit between subject and object, the subject and his/her environment. As we will see, it is also linked to the idea of going beyond the identity of a given state, hence the idea of history and evolution. I introduced this neologism in *Le sauvage et l'artifice. Les Japonais devant la nature*, Paris, Gallimard, 1986. In Japanese I translated it as *tsûtai* (通態), where 通 (to cross) also introduces the idea of going to-and-fro (in the *kayou* reading), which here is present in the ambivalent term of *ek-sist* (to stand outside).

17 On the possibility of rethinking the legal notion of ownership on mesological bases, see Sarah Vanuxem's thesis, cited in § 27.

18 Jean-Michel HIRT, La psychanalyse, entre athéisme freudien et écoute de l'événement intérieur du sujet, in Marie BALMARY et al., *Le Sacré, cet obscur objet du désir?* Paris, Albin Michel, 2009, pp. 81–116, p. 103.

19 These briefly mentioned ideas are presented and referenced in greater detail in *Écoumène, op. cit.*

20 Note of translator: the author plays here with an image similar to that of his quotation (in his previous *Histoire de l'habitat idéal*) of Manfred Clynes (himself quoted in *The Cyborg Handbook*, Routledge, 1995, p. 47):

> I thought it would be good to have a new concept, a concept of persons who can free themselves from the constraints of the environment to the extent that they wished. And I coined this word cyborg. ... The main idea was to liberate man ... to give him the bodily freedom to exist in other parts of the universe without the constraints that having evolved on earth made him subject to,

both images meaning the TOM's abstraction of any link with a place on Earth, as already symbolized by the Cartesian *cogito*.

21 *Templum* (hence the English *temple*)comes from the same root *tem-* as the Greek *temnein*, to cut. Priests with their sacred staff, symbolically cut out a quadrangle from the sky and folded it over the Earth. On these foundation rites, see Alexandre-Ph. LAGOPOULOS, *Urbanisme et sémiotique dans les sociétés préindustrielles*, Paris, Anthropos, 1995.

22 On this theme see my *Histoire de l'habitat idéal, de l'Orient vers l'Occident*, Paris, Le Félin, 2010.

23 有名万物之母, *Laozi*, I (p. 5 in the OGAWA Kanju edition, Tokyo, Chûkô bunko, 1973).

24 As I already mentioned in *Écoumène, op. cit.* p. 210, this strange expression was inspired by Chinese grammar; see MA Zhen, *Jianming shiyong hanyu yufa* (Practical handbook of Chinese grammar), Beijing, Beijingdaxue chubanshe, 1981, featuring "sentences with predicate subjects" (*zhuwei weiyu ju* 主谓谓语句), for instance. *Neige ren zui dadade* 那個人嘴大大的 (this man has a large mouth), where *zui* (mouth) is both the predicate of *ren* (man) and the subject of *da* (large, to be large). In Japanese this becomes *ano hito wa kuchi ga ookii*, where *ano hito* (that man) is the logical subject, and *kuchi* (mouth), the logical predicate, is the grammatical subject of *ookii* (large, to be large). None of this can be translated literally into French, where the sentence needs to be restructured, with just one subject that is both logical and grammatical (that man) and one predicate to include the direct object (mouth). We will come back to this point in § 9.

25 We could call this trajective time historical time, as in the Heideggerian *Geschichtlichkeit*, but it is just concrete history.

26 We will return to this problem in § 25.

27 He was asked by the author about this point during his seminar at the EHESS, on "The tetralemma in Nagarjuna" [note of translator: c. 150–c. 250 CE one of the most important Buddhist philosophers].

28 Part 3 will discuss this point in greater detail.

29 *Objectal* means: instituted as object (S) by abstraction, in other words, *objectified*. *Objective* on the other hand is the concrete reality (S/P), *objectivated* and not *objectified*. On this point, see § 13.

30 V. Sanae TSUJI, *Les impressifs japonais*, Lyon, Presses Universitaires de Lyon, 2003. In the present language the number of such impressives would be more than 2000, between 400 and 700 of them used in daily speech.

31 NĀGĀRJUNA, *Madhyamaka-karikās*, 2, 8. The verbal root *gam* (hence *gantā* and *gacchati*) is derived from the same Indo-European origin as the English *go* and the German *gehen*; hence the usual translations *a goer does not go* and *der Geher geht nicht*. Guy Bugault's translation from Sanskrit (*Stances du milieu par excellence*, Paris, Gallimard, 2002, p. 61) gives: "he who walks does not walk" (*celui qui marche ne marche pas*) and Georges Driessens' from the Tibetan (*Traité du milieu*, Paris, Seuil, 1995, p. 47): "the agent of movement does not move" (*l'agent de mouvement ne se meut pas*). We will return to these questions in § 33.

32 See Augustin BERQUE, Point de parole et paysage dans le haïku, *Revue des sciences humaines*, 282, 2006/2, pp. 29–40.

33 George LAKOFF and Mark JOHNSON, *Philosophy in the flesh*, New York, Basic Books, 1999.

34 YANABU Akira, *Kindai nihongo no shisô*, Tokyo, Hôsei Daigaku Shuppankyoku, 2004.

35 This particle is pronounced *wa* but written as (は) to distinguish it from the exclamative particle *wa* (わ).

36 *"Shugo" wa honyaku de tsukurareta*, the title of the first chapter.

37 In the original: 日本帝国ハ万世分割スヘカラサル世襲君主国トス. The final *to su* that I translated as "it is established that", is the principal, but the verb *su* (to do) has no subject. Actually, it is the same thing as our impersonal "it". When you say "it rains" it is only a formal subject. In reality there is only the predicate "rains"; but in English, unlike Japanese, it cannot stand alone.

38 In Japanese schools, the national language is taught as *kokugo* (national language) and not as *nihongo* (Japanese); Japanese dictionaries are "dictionaries of the national language".

39 Quoted from the Japanese translation by Yanabu, p. 15.

40 *Op. cit.*, p. 18.

41 Combining two sinographs, Japanese uses two phonetic syllabaries: *hiragana* (for ordinary speech) and *katakana* (for extraordinary): foreign or scientific words, or words that need to be emphasized.

42 A more elaborate version gives 我思考，故我存在, which is better but does not solve the problem.

43 We have seen above that Japanese is a Ural-Altaic language, while Chinese is a Sino-Tibetan language.

44 According to the context, this sentence could mean something quite different: "This is Descartes" (understood as for example, "the philosopher I study". Japanese grammarians delight in examples such as *Boku wa sushi desu*, which does not mean "I am sushi", but "For me, it is sushi (that I choose)".

45 This problem is the focus of a recent work ASARI Makoto's, *Nihongo to Nihon shisô* (Japanese and Japanese thought), Tokyo, Fujiwara shoten, 2008.

46 KOBAYASHI Toshiaki, *"Shutai" no yukue* (The fate of the "shutai"), Tokyo, Kôdansha, 2010, p. 61.

47 滅私奉公, where one finds the 私 also read as *watakushi*. On this history, see YASUNAGA Toshinobu, *Nihon ni okeru ooyake to watakushi* (Public and private in Japan), Tokyo, Nihon Keizai Shinbun-sha, 1976. I discuss it in *Le Sens de l'espace au Japon*, Paris, Arguments, 2004, § 38.

48 The 1887–1896 decade.

49 The 1935–1944 decade.

50 WATSUJI Tetsurô, *Fûdo, le milieu humain*, Paris, CNRS, 2011 (1935). The passage in question is quoted in § 20.

51 *Jutsugo no ronri* 述語の論理, also called *basho no ronri* 場所の論理.

52 On this question, see Augustin BERQUE (ed.), *Logique du lieu et dépassement de la modernité*, Brussels, Ousia, 2000, 2 vols.

53 Passages quoted by Kobayashi, *op. cit.* p. 201 and 202. I translate *shutaiteki* by "subjectal" to distinguish it from "subjective" (*shukanteki*), which would not be relevant to the problem.

54 On June 18, 1960 (Oku, 16 years old, was still a high school student) a huge demonstration took place against the renewal of the Japanese–American security treaty. The "struggle against the treaty" (*Anpo tôsô*) inflamed the student world from 1959 to 1970.

55 On July 2, 1964, concerning the direction of the autonomous Council of the students of Waseda, the Chûkakuha, Kaihôha and Kôkaiha currents had violent confrontations with the Kakumaruha current.

56 This term literally means "floating world". This notion comes from Buddhism where it expresses the idea of impermanence but it became known in Europe via the "images of the floating world", the *ukiyoe* 浮世絵, i.e. the prints showing various aspects of Japanese life.

57 近思録, where one finds in the chapter 治法類 (Government rights and systems) the following expression: *xiangmin wei shehui* 郷民為社会, "the villagers gather to celebrate the gods of the soil". In *shehui* 社会, 会 means "assembly, to

gather", and 社 "altar of the gods of the soil". This sinograph associates 土 "Earth" to the key 示 "ancestral altar". An anachronistic reading of the preceding quotation would give: "the villagers make society".

58 Émile DURKHEIM, *De la division du travail social*, Paris, PUF, 1998 (1893), p. 263.

59 In number 3119, October 24–30, 2009, p. 113. [Note of translator: *Telerama* is a weekly French cultural and TV magazine, published in Paris, France.]

60 *A fortiori*, according to the frequent misunderstanding of what he says in this regard. In fact, "The *bellum omnium contra omnes* (the famous war of all against all) in Thomas Hobbes is characteristic of the *state of nature* preceding the founding pact of civil society" (Patrick TORT, in Anton PANNEHOEK and Patrick TORT, *Darwinisme et marxisme*, Paris, Arkhê, 2011, p. 20). Nevertheless, it is what Durkheim denounces in Spencer: the precedence of the individual over society, which results from a pact between individuals. On the contrary, for Durkheim what is natural is not individualism, but human sociality. This is also a postulate of mesology as we will see in the second part.

61 Franck FISCHBACH, *La privation de monde. Temps, espace et capital*, Paris, Vrin, 2011, p. 60.

62 *Op. cit*., pp. 61–62.

63 I have discussed this question in *Histoire de l'habitat idéal*, *op. cit* in § 8.

64 Alain SUPIOT, L'inscription territoriale des lois, *Esprit*, November 2008, p. 164.

65 Note of translator: *fora!*, which means "(get) out!", addressed to continental French, is a nationalist motto in Corsica (which France took from Genoa in 1768).

66 Note of translator: they legalized contraception in France in 1967 and abortion in 1975.

3 Destinies of the objects

§ 13. The objectified world

The modern ontological *topos* (MOT) is not limited to the subject alone. The emergence of the modern subject has at its exact counterpart, the modern object, also firmly planted in its MOT. The one does not go without the other, as both are MOTs. This interrelation is what makes the MOT an intolerable abstraction: the so-called site-less being of the *cogito* in fact exists necessarily in a *chôra* (the medial body) to which things also belong; but the subject separates things from their *chôra*, and makes them into objects. As a subject cut off from things, the MOT is inauthentic because its establishment as such is also inextricably linked to the object as such; similarly, the objectification of the milieu into an objectal expanse is also inauthentic because it forecloses the reality of their shared links.

To be sure, "pure science" to use Descartes' term, demands such abstraction. It needs to be objective; however when it becomes scientism and legislates about things (S/P) while it only knows objects (S), it is just as irrational as the mythologies it relegated to folklore. Its world is no longer objective; it is only objectural, unduly abstracted from the reality (S/P) of our existence. This is no longer objectivation, but objectification that, in touching subjects as well as objects, presumably abolishes any subjectity, including that of humans. This is the mechanistic ideal.

Mechanicism also takes on the name of *naturalism*, because it pretends to understand what nature is. However, since this "nature" is that of the *Umgebung* (S) it is as truncated as the MOT – cut off from reality, which is not limited to the object (S) but is made up of things (S/P) that we must understand objectively as such.

– Why "must" we understand (S/P) in this way? Isn't truth posited of the object alone?

– No, that is only an inferior truth, that leaves aside our existence although we do exist and we are the ones talking about it. The superior truth mesology seeks *must* take into account this fact which goes beyond the MOT of the object. This is not just about machines, but is much more complex because it always *also* includes something of the subject. We will see in Part II that a particular logic is required that is simpler than that of mechanicism which rests on the identity principle, like engines that require repetition of the same, without which they break down. To be sure, we also need engines that run, but *above all* we need reality to remain ours. It has to remain our milieu that includes and uses mechanisms but is not limited to them.

On the other hand in the objectified world of the MOT, the link weaving things into our existence is by definition cut. What remains is the object under the abstract gaze of the *cogito* which takes itself for universal when in fact it is only contingent; what it perceives is not an object but a thing, arising out of the singular history it shares trajectively. It is precisely because it stems from a certain history that the MOT thinks it only seizes simple objects. In this history the MOT is Aristotle's grandson; as we have seen in § 6. His *Categories* organize essences into a hierarchy to favor the "first substance" which is that of the singular object. In short, separation produces essence. It opened the door to what later became the in-itself of discrete objects, abstracted from any common world (P). Everyone is in principle S and no longer S/P. To use an image, one could evoke quantum decoherence, according to which only macroscopic objects exhibit identity. They will now only *ek-sist* outside their milieu; however this is just one phase of reality which, being trajective (§ 8), goes beyond identity.

The world considered by the MOT is the product of such disconnection. It is like capitalism which counts time and space as discrete units, e.g., the square meter and "time is money". For instance, what is important today in a city is no longer the urban design (that is, the milieu) but the square meter of floors in buildings that only exist in their MOT, every man for himself. In the universal space of profit, living-together no longer means anything; it is the Tenex, Jean-Louis Curtis[1] would say, the outer darkness outside the MOT. It has been dumped there precisely because it exceeds the MOT. Its logic includes a third, which the MOT's excludes: either A or not-A, but nothing medial; A and not-A have nothing in common, that is, there is no relation between one MOT and another MOT.

As a result, the MOT feels quite lonely; deprived of its medial body, it suffers structurally from an inextinguishable lack-of-being. It is condemned to desire, the source of all suffering according to Buddhism. It is condemned to consume more and more objects because it no longer *ek-sists* in anything. In other words, ontologically speaking, the MOT is the best ally of the market, whose invisible hand is in reality its own, but foreclosed in the Tenex.

Indeed, while things transformed into objects lose their meaning, nevertheless they don't lose their market value. On the contrary, since the MOT fethishizes them ipso facto by abstracting them from their milieu. Fetishism of the self (§ 12), fetishism of the object: both go together for the same reason. Foreclosed in the darkness outside the MOT, relations that form the symbolic, ecological and technical reality of the medial body are hypostatized in the material object, which in the eyes of the MOT is all that exists. Since the physical object is the logical subject, this amounts to reducing the S/P to S (from the thing to the object) which we saw earlier in the discussion of trajection (§ 8). In other words fetishism consists in forgetting that things (S/P) only exist for us in as far as they belong to our milieu (S/P), as they are predicated by our predicative world (P), and thus regarding objects (S) as existing in themselves.

– Are you saying that fetishization is the same thing as trajection?

– No, it is only a phase, the hypostasis of P into S, which remains stuck there by *focusing on the object* (§ 14) while the historical process of trajection is an indefinite series of subsumptions of S into P, hypostases of P into S′, etc. (§ 8). Fetishism stops trajection; that makes a fetishism of S, as in the case of the MOT, a fiction based on foreclosing the milieu.

If the lack-of-being of the MOT forces it to consume more and more objects, it is not only because it unconsciously aspires to find again the medial body it foreclosed; it is because it fetishizes those objects. Hence the political economy Marx envisaged with the *1844 Manuscripts*; but even more profoundly – and ontologically – it is because for the MOT, *the possession of the fetish is the same as its essence.* Take for instance any intellectual: his compulsion to *buy* books of which he will only read a fraction is the hypostasis (S) of the knowledge (S/P) he would like to personify, to *be* a real scholar. Imelda Marcos' three thousand pairs of shoes are the hypostasis of the fleet-footed young woman she would still have liked *to be*. This is how value seeps into merchandise in the objectified world of the MOT: it wants to be what it cannot be. At least one thing is certain, it keeps the market going.

To be sure, this not only concerns merchandise. To reify representations, Marx would have said, is true of the trajection of any reality. It is what *constitutes* all civilizations with their material impact on the planet. Three thousand years of ever increasing hypostases of the Arcadian myth ends up creating urban sprawl, the Porsche Cayenne and global warming,[2] as well as the "Pythagorean mysticism" denounced by Paul Jorion,[3] in other words the fetishization of the mathematical model that finally produces the whole mechanism of our System without which we could not live. We will see in Part III that hypostases also prevail in nature and embody life itself. Obviously then we cannot just curse this fetishization! We must understand that it is nefarious in as far as it freezes the trajective process in only one of its two phases, thus producing fetishes that hide reality.

§ 14. The focus on objects

In one of the novels of Alain Robbe-Grillet (1922–2008), the pope of the "nouveau roman", a piece of tomato is foregrounded:

> A quarter tomato, truly without defect, machine sliced from a perfectly symmetrical piece of fruit.
>
> The compact and homogeneous peripheral flesh of a pretty chemical-red color has a regular thickness between a strip of shiny skin and the compartment where the seeds are stored; they are yellow, nicely calibrated, and kept in place by a thin layer of greenish jelly along the central bulge. …
>
> Right on top, a barely noticeable accident occurred: a piece of peel, detaching itself from the flesh by one or two millimeters, rises up imperceptibly.[4]

In half a page of this pretend detective novel, *The Erasers*, the pretend story stops at a piece of tomato. Do we know *why*? No, but as for *how*, we know that much: "sliced by the machine". This example of mechanical perfection ends up, purposely, producing a text without any possible interest. Like the tomato itself, any subjective concern has been cut. The only thing left is this emblematic figure of the focus on the modern object: "a quarter tomato truly without defect" is devoid of any history or milieu.

Of course, it is a game that Robbe-Grillet admits could still have a history: "a barely noticeable accident occurred", etc. Without this accident, we would only have the substance of the quarter tomato (S); but the hint of possibly saying something about it (P) quickly disappears: the passage is closed and we will not hear anything more about it. The quarter tomato remains an object (S).

These forms without history, which do not inspire painters or writers, have been replicated by modernity in mass production. Mechanically. This is the case in particular of our large high-rise blocks: their geometry which really is without defect, refuses any possible space for an accidental opening onto a world (P). No Utrillo here! They exist separately, outside any poetry or urban coherence, simply juxtaposed with the city. To be sure, because we exist, those solipsistic forms symbolize nevertheless; however what they symbolizes is precisely the cutting off of world, the *cosmo-tomy* engendering the MOT. To cut history, to cut the world into quarters is effectively the same thing: to reduce a thing to an object, the *Umwelt* to *Umgebung*, becoming to identity.

The focus on the object puts an end to the deployment of being – the movement whereby being, by *ek-sisting*, escapes from the shackles of physical identity. Why escape from identity? *Because* we are alive, and *in order to* remain alive. Only in as far as we are alive will the form of things exceed the identity of their limits (their *topos*). As Heidegger would have said, *ihr Wesen beginnt* (their being begins): their being begins together with ours. There has to be some accident enabling us to take hold, at which point a particular thing begins to exist because it concerns us. It concerns us because together we are involved in a certain history, the history of a certain milieu, a certain landscape. For example, Zong Bing, the author of the first book on landscape in human history (around 440 CE) already wrote in the very first lines: "As for the landscape while having substance it tends towards spirit."[5] In other words, the landscape is not limited to the *topos* of its material form (S); it goes beyond that and "tends" (*qu* 趣) to be something for us (S/P), trajectively.

In the case of life, mechanicism is irrelevant because of the iteration of the same. The things of life are trajective since they are not fixed objects like a pinned-down (and killed) butterfly. Didn't Haldane declare in *The Philosophy of a Biologist* (1935) that, in biology, "physics is not an exact science"?[6] This is even more valid when it concerns human life and landscapes!

§ 15. Substance or lack of substance

Let us return to the principle anticipated by Zong Bing. In its original state and its isolated substance (S) alone, the landscape is nothing like a landscape. It only becomes such (i.e., a landscape in Zong Bing's terms) or in other words its being begins as such, when a certain relation (S/P) connects it to our own existence, within a certain world (P). In mesological terms it is a trajection. In logical terms, it is a predication: S is predicated as P when there is an interpreting being. The landscape then is not an *object* (S); it is a *thing* (S/P), a trajective reality: both medial as belonging to a certain milieu and historical.

There is a confirmation: the notion of landscape appeared at certain moments in history: in the 4th century in China, in the Renaissance in Europe; and in certain milieus, of poetry in China, of painting in Europe. From then on, the idea spread historically to other milieus.

In other words, landscape is a question of *Umwelt* and not *Umgebung*. In short, it is reality (S/P), and not an in-itself (S).

The trajective moment instituting the landscape as such can be anticipated in the word *qu* 趣, used by Zong Bing to link the material (*zhi* 質) and the immaterial (*ling* 靈). The meaning of the sinograph *qu* 趣, composed of the elements 走 (to walk quickly, to run) and 取 (to take), revolves around the idea of attraction and intention. I would like to borrow it to express the relation between S and P, which would be: the environment (S) attracts us as "landscape" (P); or S/P. However this is just the first phase of trajection: S taken up into P. A second phase is missing: the hypostasis of P in S′, which concretely, in the history of real milieus, consists, e.g., in arrangements resulting in the material transformation of the landscape such as villas to enjoy the view, roads to those villas and so on.

It was normal for Zong Bing to only think of the immaterial side of that movement, given the context of his time.[7] He was influenced by Taoism and Buddhism. They command the experience of nothingness (*wu* 無) and emptiness (*kong* 空), in other words, they privilege the lack of substance over substance, unlike modern materialism. This question goes back to those we discussed in Chapter 2, especially the overturning of the modern paradigm in the Kyoto School: rendering P absolute instead of S. To be sure, this is not the case with Zong Bing because he clearly says that landscape is substance (*zhi you* 質有), *but also* (*er* 而) that it tends toward the spirit (*qu ling* 趣靈). What he does not mention however, is the second phase: while predicative, or in other words insubstantial, the landscape *also* indeed tends towards substance.

Let us translate: Zong Bing clearly anticipates the predication S/P but not the trajection (((S/P)/P′)/P″)/P‴ … etc., which concretely combines assumption and hypostasis (§ 8). Why? Because he is not a mesologist; he is a believer. In Marxian terms you could also say that he thinks about representation but not about reification, or in Husserlian terms that he thinks about noesis but not noema. As a painter, however, he could have sensed that he substantialized the landscape with his brush and his ink; but this would have been utterly outlandish in the context of the contemporary thinking about such matters.[8]

There is no need to specify that on the contrary, the MOT tends to reify and hypostatize the landscape. Of course there are intellectuals who claim that it is a mirror of the mind. However, throughout our civilization there have been scientists in particular who reduce it to the *topos* of the environment, city dwellers who noemize it as secondary residences, developers who make it clink as hard cash, in short they substantialize it. In the end the focus on the object means killing the landscape, as is obvious in the normal effects of tourism and holiday resorts.

It goes without saying that mesology opts for neither the lack of substance nor substance. It does not absolutize either S or P. What it wants to grasp is the reality of the trajection that gives us, we hope, an enduring landscape.

Notes

1 *Un saint au néon*, Paris, Denoël, 1956. [Note of translator: this is a science fiction novel where the “Tenex” stands for “ténèbres extérieures”, the outer darkness.]
2 I argue this thesis in *Histoire de l'habitat idéal*, *op. cit.* in § 8. See also below, § 35. [Note of translator: the “Cayenne” is the SUV produced by Porsche.]
3 *Op. cit* in § 8.
4 *Les Gommes* (*The Erasers*), Paris, Éditions de Minuit, 1953, p. 161.
5 *Zhi yu shanshui, zhi you er qu ling* 至於山水、質有而趣靈. Nicole VANDIER-NICOLAS, *Esthétique et peinture de paysage en Chine (des origines aux Song)*, Paris, Klincksieck, 1982, p. 64, translates: “Concerning the mountains and waters, while having a material form, they tend towards the spiritual.” “Material form” is a translation of *zhi* 質 imposed by the tradition of glosses of this passage, but the word actually means substance, matter, native state, without any idea of form.

6 Quoted by Georges CANGUILHEM, *La connaissance de la vie*, Paris, Vrin, 2009 (1965), p. 195. John Burdon Sanderson Haldane (1892–1964), the British biologist, is one of the founders of population genetics. This position is even more interesting as Haldane was a great quantifier of the living.

7 For more details, see Hubert DELAHAYE, *Les premières peintures de paysage en Chine. Aspects religieux*, Paris, École française d'Extrême-Orient, 1981.

8 In Zong Bing's world, painters thought instead they were following the natural process of Dao.

4 Acosmia, or cosmicity?

§ 16. Earth, world, cosmos, universe

Henri Raynal recently published a work called *They Decided that the Universe did Not Concern Them.*[1] The author uses the term *acosmism* to denounce the "autism" of our species which feels released of any obligation to think its place in the universe and remain there. To say something similar, I prefer to use the term *acosmia* which has the advantage of evoking the Durkheimian *anomia*, the erasure of values and the resulting disorders. However, while anomia is social, acosmia concerns both the social and the natural. More specifically, it is about engaging human values in natural processes, to which mesology refuses to reduce them, contrary to naturalism (such as sociobiology or Callicles' views in the *Gorgias*[2]) while insisting on founding values in these natural processes. This is also contrary to contemporary metabasism such as in, e.g., "French theory", which would like us to float in the blue yonder without terrestrial base. [Note of translator: Professor Berque created the neologism "metabasism" to refer to theories that deny any material base to language.]

On the one hand, there is no question of mesology reducing the human to nature since all human reality is trajective and since the human subject itself is trajective as the predicate subject of itself. Trajectivity here necessarily assumes a natural foundation (S) in a predicative dimension P which cannot be reduced to S. Reality is necessarily S/P. On the other hand there cannot be any question of reducing this S/P reality to the predicate P alone. The latter can never be extracted from its foundation in S, contrary to claims of metabasism and, for different reasons, the earlier Kyoto School that adopted Nishida's view of the predicate as literally "without a basis" (*mukitei* 無基底).

If S is the logical subject, the substance or substratum, the Earth and nature, then what is P? As we have seen, for the logician P is what can be said about S. Mesology would agree with this, but there is more to it: first of all it concerns what we feel about S, what we do with it and what we think about it. The S/P reality is the link between S and P established by the senses, language and thought. In this relation, S takes on a certain meaning that is the physical direction in time and space, a bodily sensation and mental significance expressed by words.[3] The creation of meaning [*sens*] starting with S brings about the reality of our world. Nevertheless in this relation the world assumes the position of the P predicate, which is how S appears to us. It consists of all the predicates according to which we grasp S to make up the S/P reality.

An old image will illustrate this link. In one of his orations, Isocrates (436–338 BCE) uses the following formula *hê gê hapasê hê hupo to kosmô keimenê*, "the whole Earth lying under the sky".[4] *Hupo ... keimenê*: the Earth here is in the position of *hupokeimenon* (S) with respect to the sky. Let us pursue this metaphor further; the sky is in the P position and covers the Earth subject like the male covers the female.[5]

Is this just a metaphor? Not for mesology: it includes the metaphor, the mental and verbal image but goes further. It also concerns physics and biology. Because we are bipeds and subjected to gravity, we have our feet set on the Earth but our heads lifted towards the sky, not the other way around, neither arbitrarily nor in this or that direction. Feet support us on the Earth which lies there in the position of a substratum (*hupokeimenon*). Our head thinks and speaks, in other words, announces predicates, words carried to the sky by the wind. All this amalgamates us within a specific milieu, our milieu that makes us the amalgam between Earth and sky. This is how human cosmologies are born.

In Japanese, "milieu" is called *fûdo* 風土. The first sinograph means the wind, but also the customs which are our ways of living, speaking and acting. It is here in the position of the determinant. The second sinograph signifies the Earth and is here in the position of the determined. This means that on the planet Earth, the human milieu necessarily comprises a substantial ground (*do* 土: the *hupokeimenon* S). Since we exist, the milieu necessarily includes a certain way (*fû* 風: the predicate P) in which the ground is determined by our thoughts, our words and our actions, in other words by our customs; the "way" which, by the wind (風), participates in the sky's insubstantiality, *sora*, a word written as 空, the sinograph that also translated the Sanskrit *śūnyatā*: Buddhist emptiness, as we have seen in § 9. From this emerges the S/P reality, which is both substance and non-substance, in other words the trajection from one to the other and from one into the other.

We will see in Part II what kind of logic can account for this; it is not a logic of the excluded middle. Here, let us confine ourselves to more familiar images. The sky shows the Earth in a certain light that varies from moment to moment, from one season to another, from hemisphere to hemisphere. However, it retains an invariable order because the impalpable vault is also as firm as the firmament which prevents the waters above from joining the waters below (Genesis, 1: 6–8), and invariably turns around the Southern Cross or the Pole Star (from the Greek *polein*, to turn). This cosmic order supports human milieus and guarantees their reality. This explains why in Latin as well as in Greek it has a name – *kosmos* or *mundus* – that both means the sky, the world and the ordering of things by housekeeping, of the body by grooming, of manners by decorum and by ornaments, especially feminine finery, *mundus muliebris*, with the cosmicity of its cosmetics. Finery (P) is in the position of predicate (P) with respect to our body, *sôma* (S), the latter in the position of *hupokeimenon* (S), while the world (P) is in the position of predicate with respect to the Earth (S). To say that "clothes do not make the man" is a substantialist position that ignores the trajectivity (S/P) of human reality. This is why the sovereign who keeps order down here, the Son of the Sky, was designated in China by the character *wang* 王, which allies Earth (土) and sky (一) along a cosmic axis; similarly the French *roi* (king), or the *-rix* of Vercingetorix, comes from an Indo-European root *reg* found in the word *regularity* or *right*. In another example, in Aristotle's time the supreme magistrate in Crete was called *kosmos*.

The moral of the story here is at the heart of mesology: human order cannot detach itself from the natural order; *eppur* as Galileo would have said, it cannot be reduced to the latter. It is in the position of predicate with respect to nature: both free and necessarily founded in nature by a cosmological link. Why? Because it interprets nature, not nothingness. Here are the superlatives that Plato confers on the *kosmos* in the last lines of the *Timaeus*. They all express human values:

> For this our Cosmos has received the living creatures both mortal and immortal and been thereby fulfilled; it being itself a visible Living Creature embracing the visible creatures, a perceptible God made in the image of the Intelligible, most great and good and fair and perfect in its generation – even this one Heaven sole of its kind.
>
> (http://data.perseus.org/citations/urn:cts:greekLit:tlg0059.tlg031.perseus-eng1:92c)

This is a profession of faith by a human being in harmony with his *Umwelt* (S/P) and vice-versa. It is the same for all of us, but it is also true of all living beings from bacteria to each of us humans: the surrounding reality partakes of our being itself because we are predicating it into *our* world through our senses, our thoughts, our words and our actions.

– What happens then when modernity makes the world into an objectal and neutral universe, the in-itself of an *Umgebung* (S), disconnected, legally if not factually, from any human or simply living predicate?

– What happens is a *de-worlding* which tends to deprive our values of any foundation in nature and so makes them absurd. Neither the Good nor the Beautiful nor the True can be founded tautologically in themselves. They must be founded in reference to something else as Gödel has demonstrated: "One cannot construct a *p* proposition stating the consistency of an *S* system, that would have *p* itself belong to *S*".[6] In other words, contrary to the current view and its dogmas, values cannot be founded in values, morality in morality, aesthetics in aesthetics, justice in justice, the sign in the sign, nor even physics in physics! This is certainly no small problem ...[7]

§ 17. Acosmia

In *Fûdo*, Watsuji uses the concept of *jikohakkensei*,[8] one's self discovery in the surrounding milieu, in other words those founding markers that *ek-sistence* towards the world gives to being, which assures us that we are really in the world and are really ourselves. A pathological version of this discovery was made famous by one of Maupassant's novellas, *The Horla*:

> After sleeping for around forty minutes, I opened my eyes without moving, having been awakened by some confused and bizarre emotion. At first I saw nothing, then suddenly it seemed that one of the pages of the book left open on the table had just turned by itself. No puff of air had entered through the window. After about four minutes, I saw, and yes I saw, with my own eyes another page rise up and fall back on the previous one as if a finger were flipping through. My chair was empty or seemed empty; however I understood he was there, sitting in my place and reading.[9]

The narrator has become invisible to himself and panics at seeing things existing without him. Somehow he no longer has his animal body but only his medial body: they are *his* table, *his* armchair ... someone else has taken his place, the Horla, a kind of being-there who does not have a home of his own.

In Maupassant's novella, the discovery of oneself as a Horla ends in total madness. One could see it as an inverse reflection of what happens to the MOT, whose lack-of-being results on the contrary from having lost its medial body, its own world, leaving it feeling lost without landmarks. This out-of-there,[10] this out-of-place of a *cogito* that cannot even claim to have a soul, describes *acosmia*.

– Wait a minute! Are you saying we are all completely mad?

– No, but that the world eludes us. It eludes us because mechanicism made it into a machine, the System, which more and more tends to follow only its own mechanical rules, replacing the world's flavor by the sole functionality of its own functors. Fortunately, we are still here and capable of regaining control; however, not always and not always on time. An example of such bringing under control was the rejection of the diktat: "Paris has to be adapted to the automobile", or in other words to a mechanical system. Today, we think more sanely that a city must on the contrary be adapted to human life. However, the System is not always so grossly mechanistic; it can be more subtle precisely because it hides under the predicate "human life". But "human life" does not exist in itself and will only be what we, predicate subjects of ourselves, make of it.

– In short, your "System" is the expression of ourselves?

– Indeed, because it is one aspect of our medial body. It is the mechanical side of half of our being, which with the excuse of making life easier, devours our life in order to function. This is our Horla, which lives its machine life at our expense, this "motile life of what is dead in itself", *sich in sich bewegendes Leben des Toten*, as Hegel said about money and the alienation of the world of merchandise.[11]

– "The life of the dead", those are zombie stories!

– It is not just an image, it is quite concretely acosmia. Take the advertisement of the Nissan SUV Pathfinder which appeared in every magazine in 2006. This machine, destined to sell you the game of "out into nature" was touted as having video screens in the headrests. Your children would be able to play videogames or see Batman restoring order in Gotham City while you prove to nature you love it since you are driving around in it with your SUV.

– You are raving, this "You love nature? Prove it!" was an ad for the Mitsubishi SUV Pajero in 2004.

– Same thing, it is in the System and consists first of all of interposing machines between our body and the Earth, or between our body and the world, because it is as world (P) that the Earth (S) appears to us. This really is de-worlding, "deprivation of world", according to Fischbach (§ 12), *Entweltlichung*, according to Heidegger. For Fischbach, this deprivation of world is the normal condition of the modern subject:

> I claim that in reality the subject *exists to be* thought separately from the world. Therefore the subject's separation with respect to the world is not a pathological condition from which he should be removed or cured. The absence of world is not an unfortunate accident that could befall the subject but his normal state … the loss of world constitutes the subject. … Today our problem is not that the subject is negated, but on the contrary that he is made to assert himself in extravagant proportions.[12]

Indeed, the fundamental logic of modernity is to substitute for the S/P worlding trajection (S as P, Earth as world), the acosmic juxtaposition of two subjects who have nothing left in common; they make themselves absolute separately; one is the *cogito*, the other the object (i.e., the logical subject.) The first is now its own authority while the second is subject to chance or mechanical necessity. There is no longer any room for the contingency of a common history that would unite them in a common milieu.

– Are you saying that de-worlding means the end of history and the end of the milieu by the absolutizing of the subject?

– Yes, that is what it means and you only have to open your eyes to see it in the landscape or in architects' portfolios. For instance, those gigantic high-rises imagined by Le Corbusier, one of them to be built on the hills of Algiers, the other on the site of the Orsay Station (today the Orsay Museum); those are pure metaphors of his ego (S), and therefore those pure objects (S) are mechanically repeated 1,375 kilometers away as the crow flies, in other words without any relation to either the milieu or history. They are totally acosmic and devastatingly de-worlding, for both history and the milieu where they would have been established. Fortunately it did not happen in this particular case but you can see it everywhere on Earth, to different degrees.

Since by definition acosmia has no order, it cannot be grasped in a certain order but examples can be found that at the very least can be reduced to the same principle: the stubborn *ek-sistence* of the MOT outside the milieu and history that produced it and where in reality it nevertheless exists. For example:

- Item, the end of urban composition, which was the composition of forms with other forms and composition between people. From now on, each goes his/her own way, architecturally or otherwise.
- Item, the exaltation of rupture, as in the small surrealist "exquisite corpse" papers or the cacomorphy of contemporary cities.[13] The inverse can be found in the "chain poems" (*renga*)[14] or the garden walks of the Japanese tradition where the ideal, according to Watsuji, was "to maintain harmony in change".[15]
- Item, the standardization of globalization, which is the total opposite of composition and harmony in difference. This not only kills the landscape, or *shafengjing* 殺風景, to quote Li Shangyin (813–859),[16] it also kills the milieu, *sappûdo* 殺風土, as Watsuji would undoubtedly say if he were still alive.
- Item, the Brownian movement of "friends" on the Web, whose random meetings can only pertain to a stochastic process, and not to some sort of life milieu.
- Item, the progressive transformation of suburbs into juxtapositions of private commercial spaces, and malls that replace the street. The MOT, in the mocking words of the Japanese sociologist Miura Atsushi, is no longer a citizen (*shimin* 市民), he is a private person (*shimin* 私民).[17]
- Item, rising inequality in all the rich countries and between countries; as the MOT demanded and obtained legal equality, which symbolically liberated him from any medial body, he adapted quite nicely to factual inequalities. No medial body, no worries about others.
- Item, the ideal adopted by managerial techniques: to atomize collectives, divide and individualize employees, and systematically make them feel insecure and undermine their job security.[18] It is true that *divide ut*

regnes "to divide in order to reign" is an old principle but it works best with the MOT.
- Item, the dream of cloning: to repeat oneself without having to depend on anybody, that would be fantastic. It is ideal for the MOT.
- Item, even more ideal would be to make the MOT eternal, which is what the Cambridge computer biologist Aubrey de Grey is working on; for him "the two most fundamental instincts of all living beings are survival and the transmission of their DNA".[19] Since the first is clearly more important than the second, he and his wife Adeline decided not to have children. Of course one's MOT is far more comfortable without kids, i.e., without sharing one's DNA, keeping it for oneself.
- Item, symmetrically, in the Snowflake program which works on the conservation of supernumerary embryos for *in vitro* fertilization, the MOT is already present in the very primary cells: "An embryo has 46 chromosomes and is a complete human being", Mrs. Meredith confided in *Marie-Claire*; she is the mother of little Ella who "consisted of eight cells and was only three days old when she was frozen". No doubt here about ipseity, because "God showed us the way".[20]
- Item, the playful inversion of the meaning of applying marks, tattoos and piercings which in traditional societies were literally a *kosmos*; with the MOT they signify the demarcation of the animal body from the customs of the medial body, even though they ipso facto demonstrate their belonging to contemporary "tribes", analyzed by Michel Maffesoli.[21]
- Item, the oral inflation of "a bit" or "a little bit", or quotation marks in writing, or orally in Anglo sign language; this because the MOT must mark its lack of concern, its *Teilnahmlosigkeit*, as Honneth would say.[22] These are times of channel surfing, of sliding and gliding sports: one no longer engages in the world, and lives in detached contemplation, a second degree life ...

– This is normal because the MOT travels by helicopter (§ 7)! That is even better than Le Corbusier's pillars to break away from the Earth ... Now wait, stop this outpouring. Your acosmia, we can see what it is: the contrary of cosmicity. We would like to know what you mean by that and how it might be possible today.

§ 18. Cosmicity, 1: from the body to the world, and from the world to the body

Every human society necessarily interprets the world to make it into *its* world. Such an interpretation is a cosmology, namely a saying (*logos*) about the world (*kosmos*), in which, trajectively, the object of the words (S) and the words themselves (P) become one single and same reality (S/P): *the* reality of *this* world and not another one. Anthropology and history give us several examples. However these are modern or postmodern sciences that as such participate in the objectification of the *Umwelt* into the *Umgebung* and so deconstruct any *Umgebung* to conclude that in the end we only have a construction of reality. In other words we only have the *Umwelt* or even, in Nishidian manner, a pure predicate. In that sense, cosmologies would only be, always and for ever, cosmo-*logies*, simple tales, all different and essentially equivalent. This seems to me Philippe Descola's view in *Par-delà nature et culture*.[23] His "ontological square" places "naturalism" which he identifies with modernity, at the same level as totemism, animism and analogism.

However, from a mesological point of view, modernity is different from the other three ontologies specifically in depriving the subject (S) of world, while those integrate it specifically into a world and vice-versa. Correlatively, modernity grasps the object (the other S, but in principle decoupled from the first) with a cognitive and material mastery that is immeasurably greater than that of the other ontologies. The price however is a foreclosure leading to a lack-of-being that makes MOT a crippled ontology, exceptional among the other human ones. For mesology, the modern view is not equivalent to the others. It cannot be placed on the same atemporal level because it is a phase in human evolution that follows and dominates them. Its very historicity however destines it to be overtaken by other visions.

As we have seen, modernity is characterized by de-worlding, the loss of cosmicity. In order to survey good forms of cosmicity, we have to go back to times before the arrival of the modern western paradigm. We will do so at first, and then ask if some new form of cosmicity is not already replacing modern acosmia.

From a mesological point of view, a cosmology establishes a relation (*logos*) between the world (*kosmos*) and our own existence. Explicitly or implicitly it is an *onto-cosmo-logy*. Contemporary cosmology, for instance, takes the form of an astrophysics according to which the universe is in principle a pure object that reflects modern dualism, in other words our ontology. Moreover, it suffers from recurring theories that refocus that *Umgebung* on our own existence, making it once again into a *Umwelt*, unknowingly and thus irrationally. Hence, the "anthropic principle", where the effect becomes the cause and its companion, "intelligent design", the disguised form of creationism.

These are contemporary versions of the delight Plato expresses in the *Timaeus* when he sees how the *kosmos* fits his own conception of being (§ 16); the text becomes an explicit onto-cosmology. The principle is well known: absolute being (*eidos, idea*), outside time and space, projects on the milieu (*chôra*) and on history its reflection, which becomes the relative being (*genesis*). The projection of being is at the same time the organization of the world (*kosmos*). And so *genesis* and *kosmos* match each other.

Of course, by giving it my own slant, this summary mistreats a text we have never ceased interpreting for two millennia. However, without entering into details,[24] establishing such a connection between being and the world makes the *Timaeus* into the paragon of all cosmologies. Like any cosmology it is culturally marked: the transcendence of the *idea* that is distanced from the sensible world by a "separation" (*chôrismos*), will turn out to be compatible with the absolute Being of Christianity, before engendering the *cogito*.

As traditional cosmologies are all culturally marked, to chart them would be irrelevant here; Descola's book mentioned above gives us a most useful overview. Let us rather extract a principle. From the mesological point of view, any cosmology embodies some sort of a trajective movement, or in short, the taking of S into P and the hypostasis of P into S′. In this case, it means connecting the microcosm (the human body) and the macrocosm (the world): the world (S) as the body (P), the body (S) as world (P). This is in fact the double movement of existence: *ek-sistence-towards* (of the body proper towards the world) and *existence-outwards* (from the world to the body proper). In other words trajection here is a to-and-fro, both the cosmicizing of the body and the somatizing[25] of the world.

Our languages have kept traces of both carnal and cosmic dimensions of human existence. The word *femme* in French comes from the Indo-European *dhe* (to suck) which e.g., gave the Greek *tithênê*, nurse (the qualification given by Plato to *chôra*, the relative being's milieu), the English *tit* and the French *fœtus, fils/fille, foin, fenouil, félicité* … in short like a woman, the milieu or *chôra* breastfeeds the relative being (*genesis*) that delights in it (§ 32)!

Homme (man) comes from the root *khem* and gave the Greek *chthôn* (earth, hence our "autochthonous", born form the earth itself, the Latin *humus* (same meaning), producing *human, humiliate, inhume, human* and *humanity*. We are of the earth, and in the earth we will be, after standing up straight towards the sky.

According to the Bible, the first man, Adam, embodies this cosmicity. His name means "earth" in Hebrew (*'adamah*), the clay God used to fashion him; the divine inspiration gave him life and spirit (in Latin *anima*, word related to the Greek *anemos*, wind). The "inspiration" (in Hebrew *ru'ah*, in Greek *pneuma*, in Latin *spiritus*), had no objective before the animation of humus into human: "the earth was empty and formless, while the darkness covered the abyss, God's wind swirled on the waters" (Genesis, 1: 2). Humans then are the incarnation of a purpose: the cosmic alliance of earth and sky by the wind, the inspiration, the voice and the word … we are not far from St. John's prologue *En archê ên ho Logos – In principio erat Verbum* (§ 8) … (*In the beginning was the Word.*)

The anthropic principle and intelligent design are but more recent formulations of the same biblical vision: the world is formed by the human. In short, Adam is a monotheistic and anthropocentric[26] version of what the Japanese word formed by eliding and implying human subjectity (§ 6), "wind-earth", *fûdo* 風土, a human milieu (§ 16).

Interestingly, it is not in the West that the cosmic breath has played the greatest role, but in China with the notion of *qi* 気. This sinograph is originally a pictogram imitating the twists of steam from breathing or some other natural phenomenon. It implies life, that of the body itself or surrounding nature. The *qi* is both vital breath and cosmic breath. From physics to medicine and metaphysics, not to mention aesthetics, this notion runs through the history of Chinese civilization. It is a *sui generis* concept including material energy, spiritual energy and vitality, in other words, integrating matter, life and spirit. We could even say that it the homologue, a very oriental one to be sure, of what the logos has been in the Western[27] world. Isn't the word a breath?[28]

The world's structuring by *qi* is essentially cosmologic. What is more, it is cosmo-*phanic*, because it is directly visible in the landscape by feng shui.[29] Feng shui itself can be defined as an ecomancy, that transforms extension into world through the circulation of *qi* in accordance with human existence. In what is undoubtedly the best introduction to feng shui in French, Frédéric Obringer's short book *Fengshui. L'art d'habiter la terre*,[30] the first lines read:

> To seize the breathing of the mountains, to find the good marriage between earth and wind in order for the living and the dead to happily inhabit the earth, such is the goal of the Chinese art of geomancy (*feng shui*).
>
> It is a simple and strong idea while its realization is complex and uncertain. Since antiquity Chinese cities and countryside have been scrutinized, modeled and remodeled in order for the spirits of the dead to find a resting place and show great benevolence for those who are still living. At the same time, men and women have tried to build their houses, their temples and palaces so as to agree with the general organization of the universe which also benefits them …
>
> A superficial look, given certain media malarkey, could drive us to consider *fenghsui* a baseless outmoded idea which should be left to amateurs of shady exoticism. However, we would then forget the degree to which the Chinese people as a whole are impregnated by this art that for centuries, if not millennia, showed them the world, made them appreciate happiness and tolerate misfortune, no matter how critical they may sometimes have been toward it.

After having been suppressed as a superstition by Maoism, feng shui has made a powerful comeback and has even attracted much interest in the West.[31] The present circumstances favor such an opening given the ecological impasse of contemporary civilization. A recent comparative study notes that the feng shui in vogue in China since the 1970s has led to a movement in research whose central problematic takes aim at modern dualism and materialism. The *qi* happens to occupy the very pivotal point at which they separated:

> In the present situation, there is a growing demand for going beyond Western dualism, which conceives of man as an all-powerful being,

> and to substitute a monism pleading for harmony between man and nature by reconsidering the vision of the oriental world that brings together the spirit and the body. Hence the attention paid to *qi* which is its central concept.[32]

This is precisely the problem. Leaving aside the charlatanism pervading the practice of feng shui, it cannot be sufficient to repudiate modern dualism in favor of its contrary, a monism which ultimately would be reduced to some New Age mystique, or rendering P absolute instead of S. This tendency is clear in the feng shui tradition that sometimes has curious Christian echoes, cf. the expression "to drink the forms, to take the species".[33] Of course this has nothing to do with the Last Supper; it refers to classifying landscapes in recognizable types, presumably as scenes from daily life: "the lion playing with a ball", "the old fisherman throwing his net", "the spider weaving its web", "the immortal reflected in a mirror", "five tigers catching a sheep", etc.[34] Landscape then really becomes legend (*legenda*, that which must be read). Faced with the environment, the as-predicate (P) is substituted for the physical approach (aiming at S). This amounts to overturning the logic of the Aristotelian subject for the logic of the Nishidian predicate (§ 11). To be sure, in feng shui there is not only the "method of form",[35] which is visible, but also the "method of principle",[36] which is invisible. However, in spite of these magnificent compasses that act as real cosmologists or machines for showing the cosmos, feng shui is not physics; it is a radical humanization of natural phenomena. Witness this extract from a book on feng shui during the Tang era, regarding terrestrial magnetism:

> Magnetic force is tinged with the maternal principle, and the needle was made with iron by wounding it. As they respectively have maternal and filial nature, they communicate by answering each other, and tend to recover their original comprehensiveness by curing the initial wound.[37]

To be sure, this is the time of Charlemagne, but since then, feng shui has not known any Copernican revolution. Apart from details, its spirit has remained the same. In that sense it is essentially premodern and today we cannot be satisfied with it. We cannot ignore modernity in our cosmology, but must go beyond it.

§ 19. Cosmicity, 2: beyond acosmia

Here, barely a third of the way into this book, to go beyond modernity's acosmia remains a distant goal; at this point we can only place the first milestone. If any cohesive order, that is, any possible cosmicity is to emerge from my work, by definition it will come at the end of the book. Nevertheless, we need to stop here to consider this perspective.

Since everything emerges from what it is not – and vice-versa – so it is with the cosmos: the tradition tells us that it was established out of chaos. In this case, the question is to see how a new cosmicity can be born out of acosmia. Such a view presupposes that somewhere there is a hinge, a threshold or a phase, in short a limit below or beyond which, there is either order or disorder. History and anthropology show us that human worlds have indeed been structured by such limits. Without delving too far into the past or the exotic, this would be the limit between wild space and cultivated space; cultivated first by agriculture the cosmogenetic limit becomes the edge of the forest, and then cultivated by writing, the generating limit becomes the wall of the city.[38] More generally – and this is where the very notion of world (*mundus*, *kosmos*) comes from – the limit would be the distinction between what is world and what is not world, the latter therefore foul. (Note of translator: the term used by Professor Berque for "what is not world" is "immonde", which in French means foul.) Where does this distinction come from?

Remember that the first human, Adam, was made of earth and returned to earth but while alive, he was animated by divine inspiration and he stood up towards the sky. Let us consider this old image in the light of the words of one of the greatest contemporary landscape architects, Michel Corajoud: *the landscape is the place where the sky and the earth touch.*[39] Let us add these two sinographs: 開闢, read as *kaipi*, they have a double meaning: to cultivate a still virgin terrain or the initial opening of the world by the separation of the sky and the earth. Without forgetting what has been said since § 1, let us conclude with the last sentence in the *Timaeus*: the world (*kosmos*) is the sky (*ouranos*). With those sieves to which Plato compares the *chôra*,[40] let us shake this chaotic pile out of which we can begin to outline some sort of order:

NON-WORLD	HUMAN	WORLD
earth	landscape	sky
dead matter	living flesh	animating spirit
impure	purifiable	pure
substance	reality	non-substance
towards below	here	towards above
from which	now	to which
closure	threshold	opening
attachment	existence	liberation
hypostasis	trajectivity	assumption
S	S/P	P

This is just one perspective, but let us develop it further. Other perspectives are possible, depending on the meaning and the position given to "world". For instance some speak of the "underworld", what the Romans called the *inferni*, "hell". Others say "here on the lowly earth", because it is the world of the Fall. These and other views might refer to a "better world", or even heavenly world. Here I will stick to the reassuring *coincidentia oppositorum* (the coincidence of opposites) between Plato, the champion of the transcendence of absolute being, for whom *kosmos* (world) = *ouranos* (sky), and Nishida, the champion of absolute nothingness, for whom *sekai* (world) equals *mu* (nothingness), or Buddhist emptiness *kû* 空, which sinograph can also be read to mean the sky. For these two intellectual guides the world is the sky.

Of course, those images are not sufficient. This is why the three columns above have several sections that link several orders of ideas. This relation is substantial and multi-dimensional enough not to consider it as just a metaphor. There is something that makes sense, a cosmic sense that runs throughout centuries and cultures. Obviously it has to do with the very constitution of what is human, the biped with a large brain and language with words that fly. The upright stance gave him the axis, the terms and orientation essential for every human world: dry land, an intangible sky, their vertical relation during life and juncture at the end of the world, horizontal in space to the horizon and in time to the dead. This is the root of all the metaphors in all cultures: the *primary metaphor*, that founds all "primary metaphors" of our flesh, and beyond that those of our spirit.[41] On that basis this book will try to build a proposal for a new cosmicity, beyond the aporia of modernity. But first, let us invoke the memory of the most famous philosopher of the past century.

One of Heidegger's most profound metaphors is used throughout his work, especially in *The Origin of the Work of Art*. It comes significantly close to the double meaning of the sinographs we have just seen: the opening of the forest by cultivation that produces a "clearing" (*Lichtung*), and the opening of the world by the separation of the sky and the earth. It is the double metaphor of the unfolding of being – the *weltbildend* being, forming the world which is ours. Here is the famous passage:

> That towards which the work retreats, and that which such retreat brings out, we call the earth. It is what it takes back when it comes out (*das Hervorkommende-Bergende*). ... Establishing a world, the work summons the earth (*Indem das Werk eine Welt aufstellt, stellt es die Erde her*). The work carries and maintains the earth itself in the opening of the world. *The work liberates the earth for it to be an earth.*[42]

From a mesological point of view, the metaphor can be interpreted as follows. While Bourdieu mocked Heidegger[43] as a man of letters or petty-bourgeois because he only considers works of art, human achievement as such is fundamentally that which spread the human habitat over the Earth (planet, the biosphere), the ecumene (*hê oikoumenê*, the inhabited), all human milieus,[44] through the trajection of the substratum (S: the Earth) into a certain predicate (P: such and such a world). This trajection is indeed an opening of world, a *cosmophany*: the appearing of a world, by the emergence of Being at a higher level (tra, *trans*). This is what happened in the little world of modern art, when Marcel Duchamp turned a urinal into a *Fountain*. At the level of the ecumene, however, this construction of the Earth as world is not just a perspective, a private joke for snobs, nor even just a *Weltanschauung*. First of all it rests on immense physical labor, especially the Neolithic cultivation that produced our countryside. As Marx emphasized in the *1844 Manuscripts*, Hegel grasped the essence of labor when he showed that it humanized nature, objectified the subject into the object and at the same time produced the human subject historically through the objectivity of his creation.[45] In mesological terms this is a trajection; actually *the* essential trajection that generates the human being and his milieu, the ecumene. Remember that the word *rural* has an Indo-European root *REUOS*, which means empty space. The same root gave the German *Raum*, space. The Heideggerian idea of "spatiation", *Räumung*, is probably derived from that image and goes with the *Lichtung*. However, taking S as P, the Earth as a world, never suppresses the Earth: the opening, cultivation, the appearing (*hervorkommen*) of S as P is only possible *because* S forever hides (*birgt*) an in-itself which closes itself to P *because* it shows itself as such a P. To be world, the world needs this otherness because it is founded in it. Between S and P there is always a "struggle" (*Streit*) that Heidegger sees between the Earth and the world. It is indefinitely *in order to be Earth* that the world frees the Earth from what it is itself, crystallized in its in-itself. The work of art is the symbol of this trajection and the ecumene its reality.

Be reassured: I do not explain what Heidegger means in order to make him into a mesologist; I am only saying that his formulation is useful for my argument, i.e., that the deliberate obscurity of his images conveniently allows us to translate them into mesological terms, giving the trajective process of assumptions and hypostases a cosmic dimension, which self-reflexively establishes the reality of human milieus. In short what he calls *Streit* is the equivalent of the diagonal / of the link between S and P. If you go back to the three preceding columns, you can see that it is a landscape. Doesn't this mean that cosmicity, like acosmia, is what is at stake in the landscape?[46]

Notes

1 Paris, Klincksieck, 2012.

2 Concerning justice, Callicles argued against Socrates that in the city "*as in nature*", the strong must rule the weak. Might is right! Sociobiology wants to reduce the social to the biological.

3 Note of translator: in French, these three aspects (direction, feeling, meaning) can be expressed by the same word *sens*.

4 Referenced *Oratores attici* 78 by the *Dictionnaire grec-français* of Antoine Bailly, Paris, Hachette, 1950, entry *kosmos*.

5 In medieval philosophy, the subject is the female and the predicate male. See Alain de Lille (1120–1202 or 1203), *La plainte de Natura* (*De planctu naturae*), translation from Latin and commentary by Yves Delègue, Grenoble, Jérôme Millon, 2013, p. 83: "Among those who apply Venus' grammar … [there] are some that, taking on the function of subject, ignore the predicate, while others, using the predicate, do not submit to the legal subordination of the term 'subject'", and *ibid.*, note 2:

> to the couple *subjectum / praedicatum*, used at first by grammarians, is substituted the couple *suppositum / appositum*. … Alain is neither a grammarian nor a logician, and as poet he plays with those technical terms as he pleases. The text determines that the predicate designate the male element of the couple and that the subject be the female element (which is "placed under" the "legal" position of sexual relations (*subjecti termini subjectionem legitimam*).

See also p. 105, note 1: "The subject (*suppositum*) in any sentence has the value of the feminine because it is 'placed underneath', while the predicate that is joined to it (*appositum*) has the value of the masculine." Obviously, Alain alludes to the so-called "natural" position of coitus. The predicate is masculine because with its semen it allows the feminine substantive to procreate meaning. See also § 44.

6 I use here the simplified formulation of Gödel's theorems by Jean-François GAUTIER, *L'Univers existe-t-il?*, Arles, Actes Sud, 1994, p. 146.

7 From the perspective of philosophy of science, the size of this problem can be measured with the help of Pierre KERSZBERG, *La science dans le monde de la vie*, Grenoble, Millon, 2012.

8 自己発見性, literally "self (*jiko*) discovery (*hakken*) -ness (*sei*)".

9 Guy de MAUPASSANT, *Contes et nouvelles*, vol. II, Paris, Gallimard, 1979, pp. 931–932. It is the second version of *The Horla* (1887).

10 Note of translator: "Horla" is homophonous with "hors-là", *out-of-there*, the contrary of *Dasein* (being-there).

11 In *Realphilosophie*, I, p. 240. Cited in Kostas PAPAIOANNOU, *Hegel et Marx. L'interminable débat*, Paris, Allia, 1999, pp. 9–10.

12 *Op. cit.* in § 12, pp. 32–34.

13 *Cacomorphie*, a word created on the model of *cacophonie*, where *morphê* (form) is substituted for *phônê* (voice) after *kakê* (bad).

14 In the *renga* 連歌, a poetic genre that flourished in the 15th and 16th centuries that produced the haïku, several people, up to several dozens, continue, one after the other, the verses composed by the preceding person. The essential difference with the surrealist papers is that the preceding part is not hidden: the succession does not occur in the absurd clash between unrelated points of view, but in the sequence (*ren* 連) of successive subjectities. See Earl MINER, *Japanese Linked Poetry*, Princeton, NJ: Princeton University Press, 1979.

15 Utsuri-kawari tsutsu chôwa wo tamotsu 移り変わりつつ調和を保つ, p. 260 in *Fûdo. Le milieu humain*, Paris, CNRS, 2011 (1935).

16 He used the term (in written Japanese *sappûkei*) he created, in the sense of "in the bad taste of boors". I use it here more generally in the sense of the sensible aspect of acosmia.

17 MIURA Atsushi, *Fast-fûdoka suru Nippon* (The fûdo becomes a McDonald's. Pathology of suburbanization), Tokyo, Yôsensha, 2004, p. 187.

18 Danielle LINHART, *Travailler sans les autres*, Paris, Seuil, 2009. The author's thesis is that the demand of autonomy against Taylorism has become a trap since management has appropriated this ideal from the 1968 movement to render employees insecure and thus in a kind of blackmail extract more profit from their labor.

19 Interviewed in the *Courrier international*, 806, April 13–19, 2006, p. 41, on the project SENS (Strategies for Engineered Negligible Senescence, in other words, immortality).

20 Cited by Emmanuelle EYLES, États-Unis. Nous avons adopté un embryon congelé, *Marie-Claire*, 641, January 2006, pp. 64 and 68.

21 Michel MAFFESOLI, *Le temps des tribus*, Paris, Méridiens-Klincksieck, 1988. The subtitle of this book is: "The decline of individualism in postmodern societies". At first glance this seems to contradict what I am saying; however from a mesological point of view, the MOT cannot just *ek-sist-out-of*; in practice it always and necessarily *ek-sists-towards* a certain world (P′, here a certain tribe), which it creates in as far as it rejects another (P, here "the society" loathed by the 1968 movement), in a self-reflexivity that is trajection (§ 7).

22 Axel HONNETH, *La réification. Petit traité de théorie critique*, Paris, Gallimard, 2007 (*Verdinglichung*, 2005). Following Lukács, who introduced the concept of *Verdinglichung*, the author believes that

> in the ever expanding sphere of market exchange, subjects are forced to behave as distant observers with respect to social life, as opposed to active participants because all the calculations they make during these actions, concerning what they could obtain from each other, demand a purely rational position, that would be as devoid of emotion as possible.
>
> (p. 26)

23 Paris, Gallimard, 2005.

24 I give more details in La *chôra* chez Platon, pp. 13–27 in Thierry PAQUOT and Chris YOUNÈS (eds.) *Espace et lieu dans la pensée occidentale*, Paris, La Découverte, 2012. See also below, § 32.

25 From the Greek *sôma*, body. This includes somatic effects, of which ethnology gives numerous examples (witchcraft etc.)

26 And of course andro-centric (at least in the second story of the Creation). Eve, on the other hand, is "the Living one" in Hebrew Hawwa(h); from *hayah*, living; cf. the feminine Arab name Hayat, "Life".

27 It might be useful to remember that for the Stoics, *logos* took on the meaning of breath (*pneuma*) animating the world, a conception that made it into the equivalent of *qi* and that influenced the Church Fathers (*pneuma* = *spiritus*).

28 See Denis VASSE, *L'arbre de la voix. La chair, les mots et le souffle: le sujet naissant*, Paris, Fayard, 2010.

29 This word is written 風水, "wind (*feng*)-water (*shui*)". The origins of feng shui are in a distant past, but it is during the Wei-Jin (220–589 CE) period that the notion appears regularly. What can be considered the first feng shui treatise, the *Book of Funerals* by Guo Pu (276–324 CE) announces:

> The breath straddling the wind is dispersed, and blocked by water, it stops. The ancients assembled it so that it could not be dispersed and halted its course. They therefore called it *fengshui*. The fengshui method consists in obtaining water first and then protecting it from the wind.

Cited in MAEBAYASHI Kiyokazu et al., *Ki no hikaku bunka. Chûgoku, Kankoku, Nippon* (Compared cultures of *qi*. China, Korea, Japan), Kyôto, Shôwadô, 2000, p. 97.

30 Arles, Philippe Picquier, 2001. Citations pp. 5 and 7.
31 I looked into this question in: Pourquoi cette vogue du *fengshui* au XXIe siècle?, pp. 149–168 in Jean-Jacques WUNENBURGER and Valentina TIRLONI (eds.) *Esthétiques de l'espace. Occident et Orient*, Paris, Mimesis, 2010.
32 MAEBAYASHI et al., *op. cit.*, p. iii.
33 *He xing qu lei* 喝形取類. If one reads 喝 in the fourth tone instead of the second, the meaning becomes "to call the forms with loud cries", but it does not change the general meaning of the expression.
34 HUANG Yongzhong, *Fûsui toshi* (Feng shui cities), Kyôto, Gakugei shuppan, 1999, p. 200, which feature similar figures.
35 *Xingfa* 形法.
36 *Lifa* 理法.
37 Cited p. 76 by HE Xiaoxin, *Fûsui tangen. Chûgoku fûsui no rekishi to jissai* (On the origins of feng shui. Histories and realities of feng shui in China), Kyôto, Jimbun shoin, 1995.
38 See my article Le rural, le sauvage, l'urbain (The rural, the wild, the urban), *Études rurales*, 187, 2011/1, pp. 51–61.
39 This is the title of his contribution to the collective work edited by François DAGOGNET, *Mort du paysage? Philosophie et esthétique du paysage* (Death of the landscape? Philosophy and aesthetics of landscape), Seyssel, Champ Vallon, 1982, pp. 37–50.
40 *Timaeus*, 52e and 53a.
41 The idea of *primary metaphors* was elaborated in Lakoff and Johnson, *Philosophy in the Flesh*, *op. cit.* in § 9.
42 Martin HEIDEGGER, *Chemins qui ne mènent nulle part*, Paris, Gallimard, 1962 (*Holzwege*, 1949), pp. 49–50. French translation by Wolfgang Brokmeier.
43 Pierre BOURDIEU, *L'ontologie politique de Martin Heidegger*, Paris, Minuit, 1988.
44 I use the term ecumene in the sense of *the relation of humanity to the terrestrial expanse*. It is to be distinguished from the traditional meaning in geography which is "inhabited part of the Earth".
45 Karl MARX, *The Philosophic and Economic Manuscripts of 1844*.
46 On this theme, see my *Thinking through Landscape*, Abingdon, Routledge, 2013.

PART II

Re-concretizing

5 Human mediance

§ 20. The structural moment of human existence

Climate and Culture (*Fûdo*, 1935, English translation 1961), by Watsuji Tesurô (1889–1960), begins as follows:

> The objective of this book is to reveal mediance as the structural moment of human existence. Thus the question here is not to know how the natural environment governs human existence. What is generally understood by the natural environment is something that has been taken out of its concrete soil, human mediance, in order to make it into an object. When one thinks of the relation between that thing and human life, the latter has itself already been objectified. Such a position consists in examining the relation between two objects; it does not concern human existence in its subjectity. On the contrary, that is for us the question. Even though medial phenomena are here constantly put into question, it is as expressions of human existence in its subjectity, and not as what is called the natural environment. I reject in advance any confusion on this point.[1]

The concept I translated as "mediance" is *fûdosei* 風土性 in Japanese. This concept corresponds to *fûdo*, a term already discussed in § 11, and which commonly means region or "local milieu". Literally then, *fûdosei* means "regionality", or "milieu-locality", in other words, the singularity or characteristic of a certain region. This meaning is close to the sense of land that Heidegger might have named *Gegendheit* and analyzed in a book that would have been called *Being and Space* rather than *Being and Time*. In the first pages of *Fûdo*, Watsuji clearly reveals his intention to write such a book and articulate it around the concept of *fûdosei*. In his introduction Watsuji writes that his intention was inspired by the numerous stopovers during the long voyage bringing him from Japan to Europe. As he arrived for his study tour in Germany in 1927, *Being and Time* had just been published, and became an additional source of inspiration. The book develops three examples of *fûdosei* that correspond to his journey: the monsoon milieu (eastern and southern Asia, Japan in particular), the desert milieu (the Middle East) and the "meadow" milieu (western Europe). Those are examples of regionality that hint at environmental determinism, even though Watsuji categorically rejects such a point of view immediately in the second sentence of his introduction.

Nevertheless, the theoretical position described in the first chapter, entitled "Fundamental theory of the milieu", and particularly the definition of the concept of *fûdosei* posited in the first sentence of the introduction as "the structural moment of human existence",[2] call for a translation of *fûdosei* that is different from "regionality". Indeed, this definition explicitly involves a dynamic structure, a "moment", i.e., a relation between two forces that combine the two halves of the same reality: human existence. To render this structure I forged the neologism "mediance", from the Latin, *medietas*, which means "half", but even more from its root *med-* because it gave us the French *milieu*, and also reflects the human milieus that constitute the central object of *Fûdo*.

Watsuji does not return to this definition of mediance in the rest of the book. However, on several occasions he speaks of the "duality" (*nijûsei* 二重性) of the human being, namely that it comprises both an individual and a collective side. This obviously corresponds to mediance, namely "*the pairing of a being and its milieu*", the latter including both a relation between humans as well as a relation between humans and their environment. Watsuji calls the relation between humans *aidagara* 間柄, which can be translated by "interrelation" or by "social body".

To render the unity of these two aspects (the social and the environmental) that constitute the human milieu, I use the expression "medial body," which I already used in § 12, and which means that the human milieu weaves relations together that are at the same time ecological, technical and symbolic, in one word, *eco-techno-symbolic*. Watsuji does not specifically analyze the milieu in those terms, but the examples he gives go entirely in that direction.

The problematic of *Fûdo* has its roots in the same term that in Japanese means a human being: *ningen* 人間. As we can see, this word consists of two sinograms人 and 間, which separately read as *hito* and *aida*. For Watsuji, *hito* represents the individual part of the human, and *aida* its relational part. Since in Japanese the determinant precedes the determinate, the term *ningen* in a sense takes on the meaning of "human interrelation". The study of this interrelation not only founds Watsuji's ethics[3] but also his study of milieus, or what he calls *fûdogaku* 風土学, mesology. For him, the relation between things necessarily involves the human interrelation. Mediance is our indissoluble relation to others and the environment.

Therefore, as we saw posited at the beginning of the introduction, human subjectity becomes the focal point that wholly animates the milieu as something different from the natural environment (*shizen kankyô*), that is, the scientific abstraction making it into an object. The milieu (*fûdo*) is not such an object; it is permeated by human subjectity, which in turn is permeated by its milieu. Mediance, defined as "the structural moment of human existence" expresses this interdependency.

From a philosophical point of view, Watsujian mesology openly positions itself in the wake of hermeneutical phenomenology. This could make it appear hostile to the requisites of the scientific approach, in as far as the latter must abstract the subject from the object in accordance with dualism. Mediance then would look like an a-scientific or, even worse, an anti-scientific concept. The same would be true of the other conceptual basis of mesology, trajection (§ 7), which by definition implies an overcoming of dualism. While Watsuji does not use the concept of trajection, what he writes about the relation between milieu and history comes close to it; for example:

> In the union of historicity and mediance, so to speak, history becomes flesh. If "spirit" is something opposed to matter, history cannot just be the self-deployment of spirit. It is only when spirit is *the subject* objectifying itself, and consequently only when it is a thing consisting of subjective flesh, that it produces history as self-deployment. ... The subjectity of flesh has as its base the spatial-temporal structure of human existence. Consequently, subjective flesh is not isolated flesh. ... The milieu was also the flesh of the human.[4]

These lines undoubtedly refer to Hegel's *Phenomenology of Spirit*, to which Watsuji wants to give the concrete texture of a specifically carnal embodiment of history in the milieu. At the same time, he anticipates the *Phenomenology of Perception* of Merleau-Ponty (1945). The essence of mediance as the expression of the subjectity of our existence would actually be the historicity–carnality of human milieus. Obviously such a notion is foreign to the focus on the object (§ 14) that characterizes modern dualism. Nevertheless, does that mean that mesology is incompatible with science?

The rest of this book will develop a response to this question, but it can be summarized in a few words: mesology is incompatible with the western modern-classical paradigm (§ 9), but not incompatible with science which consists in transcending a series of old paradigms. Mesology wants to go beyond modernity, but not by turning the paradigm over onto its flip side, as did the Kyoto School (§ 11) by opposing a form of spiritualism to materialism. On the contrary, it wants to show that spirit and matter generate each other continuously and trajectively, in the historicity–carnality not only of human milieus but of milieus in general, in other words, in the mediance of the living.

This Part II will only deal with human milieus, and leave living milieus for Part III. We will nevertheless outline now what we could consider in principle a continuity between the living and the human, in other words between evolution and history, which would go beyond modern reductionism as well as postmodern metabasism (§ 16). The objective of mesology is to re-engage natural history with human history. Such an objective has become imperative since Watsuji Tetsurô in the humanities and Jacob von Uexküll in the natural sciences arrived independently and practically simultaneously[5] at the same conclusion: *the milieu is not the environment.* As we have seen Watsuji distinguishes between *fûdo* (the human milieu) and *shizen kankyô* (the natural environment). In Uexküll, of whom we will say more in Chapter 8, the same idea is expressed in the distinction between *Umwelt* (the surrounding world characteristic of a species) and *Umgebung* (the objective environmental given). Uexküll's approach was strictly scientific and owes nothing to hermeneutical phenomenology; the contrary would be more the case.[6] Between the theses of the Japanese philosopher and the German naturalist, there is more than a coincidence: there is a structural homology that requires us to continue on the trail that their research has opened up for us.

§ 21. Mediance and the emergence of the human being

Watsuji's definition of mediance, "the structural moment of human existence", is not only foundational, but also remains extremely fruitful. However, it is obscure for non-philosophers, while instantly distasteful to naturalists who only see there sentimental humanism. Crippling indeed, if mesology is to re-anchor human milieus in living milieus, in other words the ecumene in the biosphere! This is why in practice I prefer the definition stated above (§ 20): mediance is "*the matching of a being and its milieu*". The being in question can be individual or collective, the human in particular or the living in general: the principle remains the same, namely that there is interdependence and adequacy between two terms: the being and its own milieu. This correlation is a dynamic coupling – one moment – which evolves in a certain sense and which for the being in question is full of meaning. In the human, this meaning may be elaborated as productions of the spirit, but is at first expressed in its physical constitution and in the dispositions of the flesh.

The objective is to clarify all this. It is a distant goal but we are well on our way. From an ontological perspective, trajection produces mediance. We approached this concept in relation to the idea that the human being is a "subject that is its own predicate" (§ 8). We then saw (§§ 18 and 19) that the world consists of all the predicates (P) according to which we grasp a subject (S) which is the Earth, or nature and that this relation is established in a to-and-fro (trajection) in which the worlding of the body and the somatization of the world simultaneously take place. All this is summed up in the formula of the trajective chain (((S/P)/P′)/P″)/P‴ … (§ 8), where predicative assumptions follow each other (S/P, i.e. to grasp S as P) and the hypostases of P into S′ (which is S/P), P′ into S″ (which is (S/P)/P′), etc. The moment has come to give a concrete illustration, beginning with the evolution that gave birth to the subject par excellence of each human milieu: *Homo sapiens sapiens.*

Here trajection will mean the evolving process itself according to the principle of mediance: the matching of the subject in question to the milieu, namely the relational and techno-symbolical whole, that it elaborated little by little from the raw material of its environment. In this perspective the "worlding of the body" will have been the deployment of its corporality in techno-symbolic systems of increasing extension and efficiency, and the "somatization of the world", the progressive transformation of the body in a feedback loop of the deployment. In short, there has been, correlatively, the *anthropization* of the environment by technique, the *humanization* of the environment through symbols and the *humanization* of the body in the feedback loop.

These are the terms used by mesology to appropriate the famous thesis of "exteriorization" that André Leroi-Gourhan discussed in *Le Geste et la parole*.[7] I won't go into details, having already commented on it in my *Écoumène*. Let us just remember that Leroi-Gourhan meant by the term that technique and symbol exteriorized certain initial functions of the "animal body" which thus became a "social body" consisting in technical and symbolic systems; the interaction between the two terms, the social body and the animal body, then provoked hominization. As we have seen in §12, mesology calls it the medial body (eco-techno-symbolic) rather than the social body (the techno-symbolic), since our technical and symbolic systems presuppose and influence our ecological environment; however the principle remains the same.

Whatever progress paleontology has made in the last half-century, nothing seems to invalidate Leroi-Gourhan's thesis. It corresponds exactly to the idea of mediance as the pairing of a being and its milieu, creating one another. To be sure, much remains mysterious in this mutual creation. For instance, what do we mean by the "feedback loop" from the medial body on the animal body? Furthermore, where is the "feedback" in Leroi-Gourhan's thesis if the social body were only the exteriorization of the animal body? From the perspective of mesology, there is indeed exteriorization through technique – for instance today the hand of the animal body extends to that of the robot seizing a stone on Mars – but there is also the contrary interiorization through the symbolic. This is how, through this exteriorization–interiorization, the trajective to-and-fro, the unity of the animal body and the social body is maintained, in other words mediance. It is thanks to the symbolic systems – starting with language – that the robot on Mars is physiologically present as representation in the neurons of the NASA engineer who controls it on Earth. This is only possible because the robot is a hypostasis of assumptions made by human minds in the course of a very long history. Both the robot and the engineer are indeed but singular and momentary expressions of a trajective chain that started with hominization and, before that, with life.

– In short, do we have then diastole through technique and systole through symbol? This image poses problems of scale, to say the least. Doesn't it pose more problems than it solves?

– That is quite possible, but clarifying these problems doesn't exempt us from conceptually articulating their basic principles. This is the meaning of the concepts of mediance and trajection.

– Aren't concepts just words? Aren't they P predicates that are not the S object that we want to know?

– Yes, they are predicates, but as such they are neither more nor less than any knowledge, and they are necessary for any knowledge. Not only do concepts make it possible to organize thought, but their role is analogous to that of algorithms that in the natural sciences compress the complexity of information so that it can be understood and used to formulate laws.[8] They are necessary in order not to get lost in the complexity of each problem they cover: especially since it has been demonstrated, to quote the mathematician Jean-Paul Delahaye, that "objects forming a whole are often more complex that the whole itself", which allow some cosmologists "to envisage that the universe as a whole could be very simple while what we observe appears complex".[9] So let us not be ashamed of conceptualizing!

The concepts of mediance and trajection can for instance clarify the thesis proposed by Timothy Taylor in *The Artificial Ape. How Technology Changed the Course of Human Evolution*.[10] The general idea in this book, quite similar to Leroi-Gourhan's, is that the *Homo* genus has not only been biological but also bio-technological since the beginning; the appearance of tools around 2.5 million years ago with the first sharpened pebbles[11] precedes its emergence by about 300,000 years. So it is not the larger brain that produced tools but, on the contrary, tools made the brain possible. This point of view agrees with that generalized by François Sigaut: "tools made man".[12] Similarly, as bipedalism prevented the baby from holding on to its mother's back as with monkeys, tools, in this case a hypothetical baby sling,[13] would have allowed it, as with marsipials, to pursue its development outside the uterus, especially that of its skull, more slowly than among other mammals. Etc. We will not go into detals: what counts for me is that it completely fits into a mesological scheme and not a Darwinian one. Taylor shows that Darwinism does not hold up in this respect since it does not take into account the bio-technicity of the genus *Homo*.

This also explains the paradoxical fact that the volume of our skull has tended to diminish over the last thirty thousand years; the continuous development of our medial body increasingly complements its initial functions. According to work done by Antoine Balzeau's team at the Muséum national d'histoire naturelle, since Cro-Magnon Man the volume of the human skull has on average gone from 1,550 cm^3 to 1,350 cm^3,[14] or −15%.[15] These findings go in the same direction as Imanishi Kinji's[16] hypothesis that the end of encephalization would be linked to the progress of language which allowed for the socialization and preservation of experiences; of course it can be attributed to writing even more and today to computers which supplement memory. Taylor conjectures that this tendency is to continue and that humans in the end will less and less need to be *sapiens*![17]

The animal body and the medial body seem well engaged in a trajective chain. The philosopher Bénédicte de Villiers expresses a similar idea when she speaks of "the mutual moulding of man and flint ... corporal and technical spheres constitute each other through a history that is human from the beginning".[18] As we have seen (§ 19), this idea was anticipated by Hegel and Marx concerning labor. From there and through Nishida, it was echoed by a naturalist like Imanishi who speaking of the living in general, mentioned "the subjectivation of the environment, the enviromentalization of the subject".[19] This is the meaning of trajection resulting in mediance.

– Other examples?

– Gladly. Here are some random ones:

- Item, the discovery-of-the-self in the environment that Watsuji talks about (§ 17), or in Heidegger the "understanding" (*Verstehen*) with things of the region (*Gegend*), found also in Watsuji as "self-understanding"[20] within the milieu: those are expressions of human mediance, in other words of *co-interaction* between our animal body and the medial body. We will see in Chapter 7 that this co-interaction implies a particular logic including the third or middle, in other words, it is not dualistic. Didn't Hegel anticipate this question when he spoke in his *Encyclopedia* of the sublation (*Aufhebung*) or transcendence of the self by the concept, when he returns to himself after having achieved his complete exterior objectivation in nature? This "self-understanding" could very well result in, e.g., what is called "geographic determinism": if every society spontaneously tends to see the "cause" of culture in the environment, it is because it simply tautologically perceives there its own medial body!
- Item, in *La terre et l'évolution humaine. Introduction géographique à l'histoire*,[21] Lucien Febvre's thought is organized around the double status of the human vis-à-vis its milieu: agent and/or patient. To be sure, the happy positivism of his time allowed him to conclude: "And thus, banned as object from geography, today's civilized man reappears prominently as a dominator, at the forefront, as an agent."[22] In other words, having rejected environmental determinism for possibilism, his positivism limited him to this alternative, avoiding the problematic of mediance. It is precisely our task to question this double nature of agent and patient, which is none other than the subject-predicate of the self (§ 8).
- Item, Angela Sirigu, from the Center of Cognitive Neuroscience at the University of Lyon I, regarding "the illusion of the rubber hand", refers to research showing not only that the brain rejects the real hand, but that the immune system treats it as alien. She concludes:

"This simple manipulation shows how malleable the mental perception of the 'self' is, as well as how intimate the relations are between the brain and and the rest of the body, including its antibodies."[23] This is a positive example of the retroaction of the medial body on the animal body; in other words, of the somatization of the world (§ 18).

- Item, by Roland Schaer:

> What do our bodies do? They burn sugar, absorb oxygen and reject carbon gas; in the process, they produce the heat that has become one of the signs of life. ... Then what does harnessing fire consist in? It is taking out of our bodies what is made inside them; it is externalizing one of the processes they accomplish; it is making the environment into a prosthesis.[24]

In other words, the animal body becomes exteriorized as a medial body.

- Item, Jean-Marie Schaeffer, concerning the emergence of the human genus, uses the term "exosomatic" in the sense that Leroi-Gourhan gave to the "social body" (incidentally without referencing the latter):

> From the perspective of well understood naturalism, we need to start with the theorem that, just as the sciences of exosomatic reproduction (the humanities and social sciences) do not lead us beyond biology, they do not bring us below the level of culture.[25]

Such a judgment would not be disavowed by mesology ...

- Item, the biologist Motokawa Tatsuo shows that if life expectancy in Japan doubled in one century, it was not only due to medical progress but also to the ceaseless progress of the technical environment of daily life: "the first half of life obeys the rules of the living, the second is biotechnical.[26] Today life has this double aspect."[27] In other words, this is the positive retroaction of the medial body on the animal body, which increases with age.

– Thank you, that is enough! The human being appeared by exteriorizing itself in the realization of its own milieu (S/P), which it reflects on itself, I understand! And now tell us more clearly what you understand by "animal body" and "medial body"!

§ 22. Animal body and medial body

In the framework of European thought, the oldest roots one can find in the problematic of the relation between animal body and medial body go back to the distinction (see § 6) between the Aristotelian *topos* and the

Platonic *chôra*: the animal body is the *topos* of our being, and the medial body is its *chôra*, in other words its milieu. The first is individual, the second collective or relational. Without a common conceptual origin, this is the same ontological structure, the same moment of duality we find in Watsuji's notion of *ningen* (the couple *hito/aida*) and the concept of mediance (§ 20).

Even so, there are in contemporary thought innumerable interpretations of the human being that are in fact similar, even without formulating such a concept. There are many varied approaches. In his *Anthropologie existentiale*,[28] for instance, André Piette speaks of the characteristic human faculty to be somewhere physically without being there mentally; he calls that the "minor mode" of presence. In other words, unlike the animal that is fully where it is, the human being can daydream[29] and be elsewhere, which is moreover what enabled him to go to the moon on July 20, 1969.[30] The same idea could be formulated by stating that the human has the capacity to symbolically wander in his *chôra* well beyond the bounds of his *topos*. Without forgetting the technology of transportation or computers, our medial body frees us from physical space through representation, either technical or symbolical; in the same vein, it frees us from the physical time in which our animal body lives. Eastern Asian aesthetics has systematically played with this with the *yubai* 餘白 procedure in painting, the "surplus white", which goes beyond the formal limits of the image, and calls for the imagination to be boundless; or in poetry with the "surplus consonance" *yuyun* 餘韻, beyond what is said formally. There one frees oneself willingly from the *topos*, by relying on the *chôra* – in Japanese you would say by relying on the *ma* 間, whose sinograph is the same as the *aida* 間 of *ningen* 人間. This exalts the trajectivity of things, the inverse of our focus on the object (§ 14), which fixes them in their *topos* – in other words reifies their qualitative interaction with our being.

Heidegger says the same thing much later when he speaks of "spatiation" (*Räumung*), which *ek-sistentially*, exceeds the contours of material forms:

> The limit is not where something ceases, but on the contrary, as the Greeks observed, at the point from which something begins to be (*sein Wesen beginnt*) … it follows that spaces receive their being from places and not from "space".[31]

The *ek-sistence* outside the limit of a form shatters the modern view that considers things as objects, each of them enclosed in its *topos*; however it has a very long history in eastern Asia. In Taoism as well as Buddhism, the focus is on the relation rather than on the substance. The relation pertains to the invisible which is "without material form": *wuxing* 無形, in other words insubstantial; but is precisely what makes material forms powerful, because it engages and motivates human existence, in "tension" (*qu* 趣) or in co-development with things. Hence the notion of the "Great Symbol"

Daxiang 大象, namely the Tao containing potentially all the particular forms, the ultimate reality that unites all beings. The Great Symbol thus has no form: *Daxiang wuxing* 大象無形, at least no form that can be delineated by a *topos*; it is like a force field, from which all forms draw their power. Indeed, the power of a form exceeds its contour. This principle has guided Chinese painting throughout its history.[32]

Not limiting oneself to aesthetics, one would like to compare the notion of medial body to what thinkers such as Michaël Hardt and Antonio Negri have called the "commons",[33] the privileged target of capitalism at least since the enclosures (§ 12). Slavoj Žižek recently defined it as:

> the shared substance of our social being whose privatization is a violent act against which we should resist, by violent means if necessary. Among them, we can distinguish commons of exterior nature, threathened by pollution and exploitation; the commons of our interior nature; and the commons of culture, the immediately socialized forms of "cognitive capital", first among them language, our main tool for communication and education, but also the shared infrastructures of public transportation, electricity and mail, etc.[34]

It goes without saying that for the *Homo economicus* to which modernity has reduced us,[35] such a thing as the "medial body" becomes a fantasy. It is spiritualism, not to mention spiritism. Reality is the MOT (§ 6). Everything we have seen so far that pertains to the medial body is reduced to objects manipulated by the animal body and measured as much as possible, and translated in terms of the MOT. The immateriality of systems as such becomes the materiality of objects that they themselves set in motion; and the objects of course are in opposition to ourselves, the subjects. For example, the automobile system is reduced to the physical combination of automobile objects; no question here of grasping the ontological incidence of the system, which made us into cyborgs.[36] The exteriorization of the animal body into the medial body vanishes to give way to simple material "prostheses", where possible in the literal sense, namely, what medicine actually calls prostheses. Beyond that, there is the objective environment. We might even want to consider certain elements such as our "extended phenotype" naturalizing the human with the hard-line geneticism of a Richard Dawkins.[37] For the biologist this meant that things like the hive or anthill are just as genetically ruled as are the behaviors of a bee or termite, and so are part of their respective phenotype, the latter reduced to the expression of the genotype. Thanks to another concept, the "meme" (*meme* forged along the model of *gene*, i.e., genus),[38] Dawkins subsumes human cultures under the same geneticism. In other words, we can no longer leave the animal body. Going one step further with the wild imaginings of a Heiner Mühlmann,[39] there is the theory of the CMS, i.e., the "cooperation under maximum stress" observed among the treeshrew (a proto-primate in south-east Asia). With this theory, we judge cultural phenomena by the quantity of noradrenaline secreted by the hypothalamus, and by those, correlatively, of adrenaline and cortisol found in urine in reaction to stress.

I see such theories as acute symptoms of the specifically MOT pathology, that is its stubborn *topism*, which retains of each being only its material contour. One might as well only consider the decibels of a melody, or violins without violinists, or the volumes of ink and paper of literature, without the texts or readers. The ultimate goal is to reduce every reality to a material mechanism, while abstracting away human existence. To be sure, mesology does not want to ignore the *topos* of things, even the most spiritual. If the spiritual can have a *topos* (for example the sanctuaries of religions) or if equations can have physical supports (such as paper or the buildings of the MIT), it is precisely because in the reality of the ecumene, *topos* and *chôra* go together. As soon as there is life, there is mediance. This does not put into doubt the existence of ink or cortisol; it means that as long as we exist, for us, the *topicity* of things is always accompanied by their *choresy*, in other words by their deployment or their *Räumung* (§ 7) as a medial, eco-technical-symbolic body, within which their existence engenders reality only as it is paired with ours. Choresy and topicity go together; they cannot do without one another; however, the MOT by definition sees only topicity. It is the focus on the object. To use a metaphor, like a physicist who would only consider the particle and not the wave, the methodological individualism of the MOT (§ 12) only sees the individual and not society, let alone the milieu. In short, it forecloses the medial body.

Countering this view is the less narrow and highly mesological perspective of Edgar Morin who writes for instance:

> Society is produced by interactions between individuals, but society once in place, reacts in return on the individuals and produces them. If there were no society and its culture, language, and acquired knowledge, we would not be human individuals. In other words, individuals produce society, which produces individuals. We are both products and producers.[40]

As we can see, in no way does this exclude individuation. Even while stressing the latter in the name of negative entropy, Bernard Stiegler writes:

> Individuation is essentially a process of negative entropy, … it is not individualization, but the formation of the individual in as far as he/she is always linked to other individuals, groups, with which he/she is "individual-ed" and he/she is never completely individualized, which means, in Simondon's[41] language that psychic individuation is also collective or social individuation, and that social individuation is always also the individuation of a pre-individual milieu: for example language.[42]

Entropy could be called the dilution of Being in a soup cooling down to absolute zero. At the inverse of entropy, mediance is precisely what creates individuals by their concrete interaction with things and others, not only psychically but also physiologically, as we have seen in Leroi-Gourhan's interpretation of the emergence of *Homo sapiens* (§ 21). The medial body is not the opposite of the animal body (as the object is to the subject) but its indispensable half, engaged with it in the structural moment of human existence.

As we have seen, this structural moment, mediance, sustains a double movement: projection through technique and retrojection through symbol. It is trajection, which is both *ek-sistence of the body towards the world and ek-sistence of the world towards the body*. This existential to-and-fro *reveals–conceals* both a gap full of meaning and value – an *Ent-Fernung*:[43] estrangement and de-estrangement – which simultaneously gives depth to our world and brings it to mind. Thanks to our techniques, the depth of our world today (summer of 2012) extends to 12.9 billion light years, i.e., the distance of galaxy SXDF-NB1006-2, discovered by a Japanese team with the giant telescope Subaru, at the summit of Mauna Kea (Hawaii). This is the farthest celestial body ever observed; thanks to our symbols, you have it in your head even where you are, and even now, you know that its light is just the same as that which before reaching you started its journey only just 800 million years after the Big Bang.

You may dream[44] now and even rove around in a Land Rover. There is room for everybody, even for Plato, who already told us that the *chôra* is like a dream.[45] Heidegger thinks that language opens our world. For sure, but we are not angels: without technique, that world would not be a human home (*ecumene*). And vice-versa: without the symbol, there would be no human home either. It would be only what the MOT wants to make of it: an objectal exterior forever foreclosed out of our existence and thus uncoupled from Being. Very simply, it would not exist!

Technique and symbol reveal–conceal the span of our existence: this is the meaning of mediance and trajection. It is because of the ceaseless coupling of our animal body and our medial body that in the background of life there is a human home on Earth, anthropized by technique and humanized by symbol.

§ 23. The feeling of things

It must be clear from the preceding, that the MOT cannot shed light on what links us to the things of our world and makes it possible for us to exist humanly. We will not go as far as to contrast it with "trans-modern oriental *chôra*" because that would mean the absurd rejection of the mesological coupling of the animal body and the medial body, which is universal. However it is a fact that eastern Asia (see § 22) has produced notions that can help us see further than the limits of our MOT. One of the most remarkable ideas comes from poetry – not accidentally by the way – "to entrust to things the task of saying what one feels" (*mono ni yosete omoi wo nobu*),[46] which we have already seen in § 9, and which appears as a category in the first poetry anthology in Japan, the *Man.yô-shû* (middle of the 8th century.) This principle will dominate Japanese aesthetics until today. More than a simple metaphor, (like our language of flowers for instance), it aims at opening one's heart to the phenomena of nature which then can express it: twilight will be solitude, dead leaves decay, etc. The feeling of seasons[47] occupies a privileged place; a thousand years after the *Man.yô-shû*, it will eventually lead to the codification of "seasonal words" (*kigo*) in the haiku, as we will see in Chapter 6.

Similarly, where "Milly or the native land"[48] ends with a question that has no an answer:

> Inanimate objects, do you really have a soul
> That fastens to our soul and forces it to love?

Japanese thought, far from the oxymoron (inanimate/soul) [*anima*][49] that Lamartine stumbled over, acquired a real concept: "the emotional power of things" (*mono no aware* 物の哀れ). The expression (or other related ones) can already be found in the literature of the Heian period (8th–12th century), but Motoori Norinaga (1730–1801) theorized it in the "national studies" (*kokugaku*) movements developed under the Tokugawas in reaction to the regime's neo-Confucian model. The point was to recover the authenticity that had been buried by "chinoiserie" (*karagokoro*, "the Chinese heart" or also, through homophony, "the empty heart"). National studies looked for it in ancient literature and places marked by history.

It would not be absurd to establish a parallel between that reaction and the collective German reaction against the universalism of the Enlightenment, in the 18th century. Norinaga[50] and Herder (1744–1803) are contemporaries. According to Louis Dumont, there is "factual identity" between collective identities and universalist ideology.[51] It would nevertheless be difficult to attribute "universalism" to Norinaga; on the contrary, he is a precursor of the *nihonjinron* (Japanology, nativism), that sees Japan as particularly unique. However, according to Karaki Junzô,[52] on the one hand his method consists in very orthodox philology, still quite valid in many respects, while on the other hand the concept of the emotional power of things that he advances is itself characteristically universal from the perspective of mesology; this is what interests us here.

From this perspective of course, the question is not about the universality of the object that abstracts from the fact that we exist; beyond this subterfuge, which of course is indispensable for calculative knowledge, it concerns the trajectivity of things. This means in the second phase of knowledge (sublation, *Aufhebung*) to take into account our existence, an objective fact if any. This is the answer to Lamartine's question, which dualism could only foreclose. However, since even physics, at the very forefront of dualism, commands us to understand the interrelation at stake in the most objective act of observation (§ 29), it is time for us to explore humanity's patrimony in this regard. The concept of *mono no aware* is part of it in the same way as *chôra*. It does not just pertain to Japanology or to what we could call *a Japanese form of trajectivism*, but to the reality of things in a human world.

Norinaga especially studied the notion of *mono no aware* in *Isonokami sasamegoto*. Let me quote Karaki's comment:

> Encountering things, the heart is moved. It is moved, "Ah, it is beautiful" when it sees beautiful things, the moon, snow, flowers. This emotion "Ah!" is for Norinaga the *aware*: "This word *aware* 阿波禮[53] profoundly touches the heart. Later on, it only meant sadness, which then was transcribed as 哀, but sadness is only one of the aspects of *aware*, whose meaning is not confined to that. In the *Man.yô-shû*, it could be written among others as 怜,[54] read as *aware*." "In short, one could say that *aware* also contains the meaning of *okashiki*.[55] To say it differently, among the various human emotions, the stimulating and pleasant things are not very moving. Sad things, on the contrary, move us profoundly. The depth of feeling is especially the *aware*. Even though for ordinary people it is only sadness, it actually concerns the faculty of feeling (*kokorobae*)."[56]

The word *kokorobae* is interesting; written as 心延え, it literally means, "the extension of the heart". Extension to what? To things, or to the other, outside oneself. Empathy? Not only that. *Mono no aware* has exactly the same ambivalence as one of its possible translations, "the feeling of things": grammatically, one cannot know if "*no*" or "of" here signifies an objective genitive (what one would feel about things) or a subjective genitive (what the things themselves would feel). In the framework of dualism, it can obviously only be the first alternative, while the second is absurd. The focus on the object requires us to see things only as objects that by definition are devoid of feeling. For mesology, it is different: in the concrete reality of milieus and history, things are never either properly objective (S, purely in-itself) nor properly subjective (P, pure imaginings); they are always and necessarily trajective (S/P) that is, their reality itself is marked by feelings since it involves our existence. Meticulous scientific protocols are needed to abstract them, and even so, precisely to make them into abstractions; preferably into pure objects. Concrete reality however is something else; precisely that which mesology aims to grasp. Precisely for the sake of objectivity! But a form of objectivity that is less simple than the subterfuge of dualism …

From that point of view, the things (S/P) in *our* milieu – not the objects (S) of *any* environment – are *our* medial body, and it is because of this mediance that they can move our animal body. Or, and it is the same thing, it is in them – in as far as they are trajective – that our animal body can be moved, noradrenalin and cortisol included. This is how mesology can give an account of their emotional power. More simply said, things concern our existence. This concern is the source of feeling.

What was Norinaga's thinking about this? Karaki[57] notes that he often uses the expression *koto no kokoro* 事の心, "the heart (sensibility) of things". The term *koto* seems crucial. Transcribed 事, it usually means "thing, fact, matter", but formerly it also meant "speech", transcribed as 言, which meaning subsists in *kotoba*, "language, tongue, speech, word" (the transcription 言葉 literally means "leaf of speech"[58]). Of course, words move us. Then how do we go from thing to word, *mono kara koto e* 物から事.言へ? For mesology, it is through trajection: the thing S is taken as the word P, that is the reality S/P. Obviously Norinaga does not pose the question in those terms. For him what intervenes is divine action, *kami no mi-shiwaza* 神の御所為, the common origins of human feelings as well as natural phenomena. This common origin is the source of the phenomena that move us and of the possibility of *koto no kokoro*.

Parallels could be established with the *Naturphilosophie* of Romanticism. For mesology, the reason for these things is the mediance of our milieu. We will come back later (§ 25) to the relation to divinity. Let us stick with the idea that Norinaga, in terms of his own time, put his finger on a universal idea: in a human milieu, things can move us.

§ 24. Being towards life and practical morality

Since our medial body – as distinctively human rather than as simply living – is specifically the exteriorization of our animal body into a social body constituted by technical and symbolic systems, it is self-evident that the social body is not subject to the mortality of the individual animal body. Technique survives the artisan; the poem survives the poet; the city survives the inhabitants; language, those who speak it; and the nation survives those who die for their country. In short the medial being survives the individual being. Horace understood this when he wrote those famous verses:

> *Exegi monumentum aere perennius*
> I achieved the more durable monument than brass
> *Regalique situ pyramidum altius*
> Loftier than the regal structure of the pyramids,
> *Quod non imber edax, non Aquilo inpotens*
> Which neither the corroding shower nor the powerless north wind
> *Possit diruere aut innumerabilis*
> Can destroy; no, not even unending years
> *Annorum series et fuga temporum.*
> Nor the flight of time itself.

> *Non omnis moriar*
> I shall not entirely die.[59]

In fact, you have just heard Horace (Quintus Horatius Flaccus, born in 65 BCE, died in 8 CE) through the intervention of various technical and symbolic systems: language, writing, a dictionary for translation, printing, paper mills, an author, an editor ..., in other words it is because, to some degree, you share with him the same medial body. This shows our common mediance as human beings, for these things are beyond the reach of animals. The animal returns to the ecosystem after death; so does our animal body; but our medial body survives, inasmuch as we are human. In the *Oration on the Dignity of Man* by Pico della Mirandola, doesn't God say to Adam: "I made you neither mortal nor immortal"? And in fact, the human being is *transmortal* through his mediance.

In *Being and Time*, using the terminology of *Dasein*, Heidegger never ceases to speak of what I call here the "medial body": *außer sich sein* (being outside oneself), *sein bei den Dingen* (being with the things), *in der Welt sein* (being in the world) ..., in short, of the *ek-sistence* of the human being outside the *topos* of his/her individual animal body; curiously, at the same time he speaks of *sein zum Tode*, "being towards death", which means that the being of *Dasein* has as its absolute limit the horizon of the death of his animal body. This is a flagrant contradiction – a real absurdity. Without calling attention to this contradiction as such, Watsuji nevertheless comments:

> This is where I saw the limit of Heidegger's work. Temporality without corresponding spatiality is not real temporality. In fact, Heidegger did not go any further because his *Dasein* is in the end only an individual. He apprehends human existence only as the existence of an individual human. Considering the duality of human existence, which is both individual and social, this is only an abstract aspect.[60]
>
> Historicity is the *structure of the social being*. It will show the dual character, finite–infinite of human existence just as clearly. The individual dies, relations between individuals change, but even as they die and change incessantly, individuals live and their interrelation (*aida*) continues. It is by *the fact of ending* incessantly that the latter *continues* incessantly. While from the point of view of the individual, this is "being towards death",[61] from society's point of view, it is "being towards life".[62] And thus human existence is individual–social. However, it is not only historicity that structures the social being. *Mediance also structures social being*, and is thus also indissoluble. History becomes flesh in a sense in the unity of historicity and mediance.[63]

Only the MOT can think that the human being ends with the death of the animal body. All other human societies, each in its own way, have always symbolically expressed the opposite conviction, even without knowing that it is through our mediance that our medial aspect survives our animal aspect. Many burial practices testify to this. They may have distinguished the human genus (*Homo*) even before the emergence of our species (*H. sapiens sapiens*).

According to Yves Coppens:

> The care for the dead seems to have existed since 500,000 years ago, as shown by the discovery, on the site of Atapuerca in Spain, of a well where corpses had been thrown, as well as a magnificent red axe.[64]

Taking care of the dead of course means to continue the human interrelation beyond death. It is precisely the being-towards-life to which Watsuji refers. In the course of history, this became the belief in a beyond requiring real funerals to prepare the dead for the new life. It appears already in the oldest proven burial grounds that go back more than 100,000 years (between 170,000 and 120,000 in Mugharet-el-Tabun, 100,000 in Mugharet-el-Skhul, 92,000 in Djebel-Qafzeh, in today's Israel). Those burial grounds contain skeletons of Neanderthals and modern humans. Burying the dead in real graves, often with objects destined to accompany them in the beyond undoubtedly testifies to a sense of the sacred, in other words to the capacity of establishing communication with the supernatural. Animal species, even those closest to us, have nothing like this. This shows the sacrality of being-towards-life that is specifically ours as humans.

The forms of this being-towards-life are just as diverse as the forms of human interrelation, in other words, of cultures. The most remarkable form, in that it expresses most literally the animal body/medial body mediance, is the theory of the two bodies of the king: one mortal body and a body "that lasts forever" (*quod semper est*). As Kantorowicz[65] has shown, it was established in the medieval West, from the older expression *Dignitas non moritur*, (dignity does not die.) Everyone in France knows the traditional formula "The King is dead, long live the King!" which expresses the same idea and was used for the first time in 1422 at the death of Charles VI, and led immediately to the nomination of his son Charles VII to the throne. At stake here are two different animal bodies, one of which dies while the medial body is not interrupted and remains the same: it is royalty. Another formula, "death seizes life" also signifies the continuity of the medial body: as soon as the sovereign is dead, the sovereign lives.

On a more general and properly cosmic level, the duality in the unity of the pair, animal body/medial body, is rendered metaphorically in the work of Alain de Lille (1120–1202). *The Plaint of Nature* (*De planctu naturae*) that was one of the sources of inspiration for the *Roman de la rose*. It presents the image of the two hammers that nature entrusted to Venus in order to make the world:

> To have secure instruments in order to avoid the proliferation of bad works, I [Nature] gave her [Venus] two legitimate hammers that would

> allow her to thwart the Parcae and offer to essence the multiple modes of things. I also entrusted her art with noble workshops with their anvils, telling her to grant them those hammers to faithfully form the image of things.[66]

Let us return to less profane subjects. Like the invention of the state by pontifical jurists of the 11th to 13th centuries,[67] the theory of the two bodies of the king is inspired by Christian theology, with the idea of the Church as Christ's mystical body extended to all baptized and just people of all nations, to every epoch in history. The first formulation can be found in Saint Paul's Epistle to the Romans: "For as we have many members in one body, and all members have not the same office, so we, being many, are one body in Christ, and every one members one of another" (Romans, 12: 4–5, King James Version).[68]

Even though it was applied to the instantaneity of royal succession, the formula "death seizes life" comes from private law, where it meant that "the submission of goods and rights of the deceased is granted to the heir on the very day of death".[69] Inheritance instantiates the transmissibility of the patrimony as family property, in other words as supra-individual, even if it can be appropriated individually by each particular generation. It is the expression of the continuity of the medial body. The MOT itself likes it more than anyone else in as far as it sees personal property as discrete objects, each of them fetishized in its MOT. However what is important from the mesological perspective is the institution of patrimony which embodies the "being-towards-life" of private individuals in societies where it exists.

Nevertheless, patrimony is but a minor form of being-towards-life. The major form is certainly the notion of the soul, with many other analogous notions in different societies, in other words something that specifically symbolizes life after death. Via the Latin *anima*, the French word *âme* is derived from an Indo-European root, *ani-* or *ane-*, which means the life breath. The same root has given the Greek *anemos* (wind) and words such as anemometer, animal, animate, unanimous (isn't unanimity a beautiful expression of mediance in the human group?) … After its last expiration, this life breath symbolically continues in the immortality of the soul. In the Christian world, it remains personal, leading to the resurrection of the flesh at the end of times,[70] notwithstanding the strong analogies with the divine breath that animated the clay that shaped Adam (§ 18); in other worlds, however, it can wander from one corporal identity to another. The notion of metempsychosis comes from the Greek, where the word means the "transfer (*meta*) of the soul (*psuchê*) into (*en*) [another body]". It marked orphism, which believed in the immortality of the soul, and in the cycle of reincarnations (the myth of Eurydice is related to it.) The mathematician known for the square of the hypotenuse, Pythagoras (6th century BCE) was just as well known by the ancients for the following anecdote reported by Diogenes Laertius (*Lives and Doctrines of Eminent Philosophers* VIII, 36):

> The story is that one day, when he was walking past someone who mistreated his dog, he felt pity and said to the person the following words: "Stop and don't hit any more because it is the soul of a man who was my friend, whom I recognized when hearing the sound of his voice."

However, the transmigration of souls was established most dogmatically in India. For Brahmanism the law of karma means that the individual soul (*ātman*, the word has the same root as the German *Atem*, breath, and our *atmosphere*) is reincarnated in diverse forms, depending on our actions: (*karman* comes from the root *KRI* which means act, action, rite); for sins of action, there is reincarnation in vegetal form; for sins of speech, in animal form, for sins of thought, in the form of an untouchable (well known to India).

This shows that the medial body is also expressed morally. Perhaps Freud would have seen it as an *Über-Ich*, a super-ego, which is what it actually is since it exceeds the *topos* of our individual animal body. This means that the MOT also counts among its defects the elision of the medial foundations of morality which concerns human interrelation.[71] It is structurally particularly allergic to altruism: risking absurdity (like the *Dasein* with its being-towards-death), it can only negate this drive, which shows human mediance, by blaming it, à la Dawkins, on the "egoism" of the gene.[72]

Our present civilization, founded on the MOT in spite of the human condition, is continuously exposed to such contradictions. The being-towards-death of the MOT ontologically threatens the being-towards-life that constitutes human societies, and that traditional societies always took it into account, each in its own way. We have seen above (§ 18) an example with feng shui: each generation must take care of the life breath (*qi*) transmitted by the ancestors, the upstream side, by taking care of tombs, those "houses of the shade" (*yinzhai* 陰宅) as well as as of the downstream side, the "house of the sunny side" (*yangzhai* 陽宅), so that the *qi* may benefit not only the living but also their descendants.[73]

In this case, the *qi* embodies the inter-generational responsibility that in our climes was better understood as patrimony, but which the MOT repudiates ontologically. A good example is the ineffectiveness in today's world of the *Prinzip Verantwortung*,[74] the principle of responsibility put forward in the name of "sustainable development" and of "future generations", according to the standard formula: "development that meets the needs of today's generation without compromising the capacity of future generations to meet theirs". At the time I was writing this, all of France was buzzing with the declaration of a prominent minister according to whom nuclear energy was a "technology of the future". Of the future? Indeed, since we are completely incapable of eliminating the radioactive waste that increases massively every day, and may last several hundreds of thousands of years; in other words, a duration longer than the past term of the species *H. sapiens sapiens*, from which our present generation holds its life breath. What will it do with this inheritance? Nothing, because the MOT, having foreclosed its medial body, has at the same time rejected any intergenerational responsibility. *Seien wir zum Tode!*[75] *Let us be towards death.* And to quote the Marquise de Pompadour, "après nous le déluge", after us, all hell may break loose but who cares?

§ 25. Mediance and religion

In "Double bodies", his afterword to Kantorowicz's *Les deux corps du roi* (§ 24), Alain Boureau writes:

> Which other medieval peculiarity could produce the development of corporative identity, if not the existence of the Christian Church, based on the incarnation of God? Kantorowicz indeed describes the process in the subtitle of his book: the medieval West developed a "political theology".[76]

Later on he speaks of the "immanent sacrality that constitutes the social link".[77]

This statement echoes classic Durkheimian sociology that clearly posits the double dimension of the human being, both social and individual, and considers religion as the expression of the transcendence of the social relation. For Durkheim, who clarifies this in *The Elementary Forms of the Religious Life* (1912), religion is not just the most primitive of all social phenomena, but a universal one, characteristic of humanity. It is a hypostasis of society itself that translates the power of the social relation felt obscurely by the individual.[78]

It will be easier to understand how religious forms hypostatize or substantialize the social being when considered in the light of the two basic concepts of mesology, mediance and trajection. In monotheisms, such hypostasis is more remarkable than anywhere else. In Christianity for instance, the Gospel according to St. John starts as follows: "In the beginning was the Word and the Word was with God and the Word was God." In the mesological interpretation, such an introduction not only means that this religion like all mysticisms, absolutizes its own predicates. Here "the Word", which like all words is said about something and especially about substances, explicitly hypostatizes the predicate by making it the substance or the absolute subject: God. At the same time it absolutizes the structure of the human as subject predicate of itself (§ 8). This confirms the divine incarnation in Christ, God made man, in whose flesh and blood the faithful will commune. This same religion states no less explicitly that Christ is life: Jesus tells Thomas who asks him what road to follow: "Jesus said unto him, 'I am the Way, the Truth, and the Life; no man cometh unto the Father, but by Me'" (John, 14: 6). More materially, in the form of spires, naves and bell towers, this life of a similar medial body would later carry the Earth to the sky, by the million tons.

Whether it be as the Koran or the Tjukurrpa,[79] the mystical bond absolutizing the Word or the Gospels is essentially mysterious. This mystery is sacred because it is that of being itself, in which are reflected self-referentially the predicate and the subject of our humanity. It is the mystery of being an "I" which is also a "we" and of being able to say in French, for instance, "je me", in other words to be able to reflect a medial body (language) in an animal body. A potential infinity in finitude, this symbol is truly unfathomable. We do not know for sure where it starts, nor to which beyond (*meta*) it carries us, *metapherei*. Where does this metaphor start? At least immediately with our flesh, in the naked species of *Homo sapiens*: this flesh is the only one that has hidden behind mane or hair the lips that give birth, expressing the species in the nascent individual, just like the head where personal lips speak the language of the collectivity ... It is a deliberately obscure symbol that distinguishes us even more clearly from the animals, by indefinitely replicating fig leaves and head coverings. The places where the animal and the medial embrace, the being-towards-death and the being-towards-collective life, are sacred places, and so they must be hidden. This is probably not what Hesiod had in mind when he wrote in *Works and Days* (42),

> *Krupsantes gar echousi theoi bion anthrôpoisi*
> For the Gods hid from humans the means of life.

But who will ever know? What keeps us alive, us humans, will always remain the mystery of this self reference, which comes to us from Earth (nature) but takes us to the sky beyond any scale?

In the example of monotheism, the sacred heart of mediance – the mutual engagement of the animal body and the medial body – leads effectively to transcendence. However, as the medial body by definition extends to our whole environment, the sacred too can expand and become immanence. To be sure, modern dualism has desecrated nature which has become that neutral object measured by a subject transcending his/her milieu, but all religions carved sacred domains there. Even monotheisms have sacred places. How is this possible? In *Fûdo*, Watsuji uses the concept (see § 17) of *jikohakkensei*, the uncovering of the self in the milieu. It is an effect of our *ek-sistence*. The most noteworthy aspect of mediance goes beyond what Heidegger calls "to be with things", the *ek-sistence* which belongs to the everyday. In mediance, societies discover their being in their milieu symbolically. This discovery is hypostatized in high places that become sacred. Why sacred? Because they are obscurely felt as the source of being itself. This source is the most important resource, where periodically and ritually the human is revitalized in its being-there-together, in the medial identity of animal bodies. Thus do the Amerindians Huichols with their pilgrimage to Cerro Quemado, the Muslims with the pilgrimage to Mecca; or at the end of the shogunate heralding Meiji, in Japan where the pilgrimage of Ise revitalized the social link in the cult of the imperial ancestor, the Sun goddess.

Watsuji also, in another work,[80] commenting on the individual person/collective person relation (*Einzelperson/Gesamtperson*), attributes to Scheler the idea that the relation is constituted by no other than the personal God. This is similar to what we have just stated: the gods are a symbolic form of mediance. In Christianity, it is the sacrament of the Eucharist defined by Catholic and Orthodox theology as the species (bread and wine) that substantially contain the body and blood, the soul and divinity of Jesus Christ. One could not possibly hypostatize a predicate more concretely! Bossuet understood the essence of this symbol when he wrote: "Let us partake of this sacred meal of the Eucharist, and live in unity with our brothers."[81]

What is in fact "this unity with our brothers" but the exaltation of our medial body? It is Christ in this case, but the potential unanimity among humans can also be exalted in many other ways, in particular in phenomena of conformism and mass behavior studied by Gabriel Tarde (1843–1904). Oddly, he radically rejected Durkheim's ideas; for him, as for Mrs. Thatcher, society was but a fiction, whose only reality was individuals. Nevertheless he pointed out an "inter-psychology" which turned out in practice to be the same phenomenon: the manifestation of our mediance.

Why then are humans obliged to show their mediance, occasionally or ritually? Because on the inside, the individual is structurally incomplete. We have seen (§ 13) that MOTs suffer from a lack-of-being that condemns them structurally to the desire for objects and their consumption. This is because their individualism makes them unable to recognize their mediance. Without having any idea of this concept or what it entails, traditional societies systematically did the opposite by symbolically filling in the gap and thus making their members fully human. This is the meaning of *anthropopoiesis*, defined as follows by Claude Calame commenting on one of Pindar's choral chants:[82] "the construction of man as a cultural being, as belonging to a particular culture; the fabrication of the human which occurs starting with the constituent lack of wholeness and plasticity by means of practical symbolic creation".

Religions typically embody this anthropopoiesis, whose structural moment of our mediance entails a need: the multiform gods fill a void that otherwise would remain a lack-of-being. They complete our humanity.

In other words the human being is structurally inclined to believe in some divinity. Wondering about "the need to believe" in a recent book,[83] Dany-Robert Dufour attributes it to our neoteny, that is to the fact that the human, at birth is an unfinished being. It is a *neotene*, a premature being. From Plato to Lacan, the idea of the incompletion of the human is ancient, but it is only around 1920 that it was interpreted it as neoteny,[84] a theory that

> sees man as a being with a premature birth, that is both incapable of achieving its complete germinal development but nevertheless capable of reproducing himself and transmitting the juvenile characteristics that are normally transitory in other animals. This unfinished animal,

unlike the other animals, therefore must complete itself outside its first nature, in a second nature, generally called culture.

(p. 18)

For mesology this obviously calls to mind the problematic of mediance. In his own fashion, Dufour applies it to the problem of God, more specifically to the "death of God" announced by Nietzsche. His more psychoanalytic approach develops Feuerbach's idea that "religious speculation inverts the natural order of things";[85] according to Dufour, this means that man invents God by "an inversion that presents the creature as the creator" (p. 94). The neotene creates a "Grand Other" (God) and subjects himself voluntarily to him, because "if there is no Other to refer to, he cannot think himself as subject" (p. 101). In reality, this Grand Other who "homesticates"[86] him is no other than his own double: "man is suffering from a lack of himself" (*ibid.*). The book details the multiple forms of the neotene's voluntary enslavement, and ends up asking the question: "To leave the neoteny … but to go where?" (pp. 335 ff.). Without dreaming of a superman, he opts for "a prolongation of fictions needed by neotenes" (p. 144) but not in the form of a new religion:

> We do not really need a new religion because … we already have one, a very ancient one. What links us together and so forms a religion (from the Latin *religare*, to link) is our very ancient common aptitude for discourse. This "religion" operates without clergy, only with individuals speaking in their own name. … It would be a very novel thing to simply go along with our paradoxical nature and to agree that we do not need a new god, but a logical and secular middle term.[87] … Our one remaining experience would be that of a new transcendence that each would carry inside, immanently.
>
> (pp. 344–346)

This would mean that we assume our mediance as properly human, rather than through the detour of a Grand Other. However, as Dufour glimpses when speaking fleetingly of a "logical third", such an assumption is impossible in the logical framework of the excluded third that sustained modernity and led e.g. to the de-symbolization of "the death of God". We will deal with this problem in Chapter 7; but it presupposes that we first re-concretize those things reduced to objects by dualism.

Notes

1 WATSUJI Tetsurô, *Fûdo. Le milieu humain*, Paris, CNRS, 2011 (*Fûdo*, 1935), p. 35.

2 In Japanese: *ningen sonzai no kôzô keiki* 人間存在の構造契機. Japanese philosophy directly translated from the German *Strukturmoment* the notion of *kôzo keiki*, "the structural moment".

3 Watsuji's main work consists in a monumental *Ethics* (*Rinrigaku*) in three volumes (1937–1949), announced in 1934 by a book called *Ethics as the Study of the Human Interrelation* (*Ningen no gaku toshite no rinrigaku*).

4 *Fûdo, op. cit.*, pp. 50–52. What I translated by "flesh" is *nikutai* 肉体 in Watsuji's text.

5 *Fûdo*, consisting of a series of articles written from 1928 on, appears in 1935. In 1934, Uexküll publishes *Streifzüge durch die Umwelten von Tieren und Menschen* (Raids into the ambient worlds of animals and humans), translated in French entitled *Mondes animaux et monde humain* (Denoël, 1965) and *Milieu animal et milieu humain* (Payot & Rivages, 2010).

6 Heidegger was very interested in Uexküll's thought, and even devoted a seminar to him in 1929–1930. Giorgio AGAMBEN, *L'Ouvert. De l'homme et de l'animal*, Paris, Payot & Rivages, 2002, insists on this influence. Heidegger's own course was published after his death: *Les concepts fondamentaux de la métaphysique. Monde–finitude–solitude*, Paris, Gallimard, 1992 (1983).

7 Paris, Albin Michel, 1964, 2 vols. English translation: *Gesture and Speech*, Cambridge, MA, MIT Press, 1993.

8 See Gregory J. CHAITIN, *Algorithmic Information Theory*, Cambridge, Cambridge University Press, 1987. This is how, for instance, Kepler's laws "compressed" observations made by Tycho Brahe.

9 Jean-Paul DELAHAYE, *Complexité aléatoire et complexité organisée*, Versailles, Quae, 2009, pp. 53–54. "Simplicity" here needs to be seen in the framework of "the complexity of Kolmogorov", namely in the sense of an object that can be defined by a short formula, as for the numbers of π, whose series reaching today 1,240 billion nevertheless gives the illusion of randomness (p. 51).

10 Basingstoke, Palgrave Macmillan, 2010.

11 Unlike accidental "tools" that can be used by many animals to accomplish certains tasks, a real tool presupposes a specific elaboration and the retention of the thing for repeated use.

12 François SIGAUT, *Comment* Homo *devint* faber. *Comment l'outil fit l'homme*, Paris, CNRS éditions, 2012.

13 Taylor posits, quite plausibly, that the material of this baby sling (leather or vegetal) did not allow its preservation.

14 Between birth and adulthood, the brain's volume is multiplied by 4 in humans, as opposed to 2.5 in chimpanzees.

15 *Science & vie*, April 2011, p. 20. this evolution is accompanied by various local modifications; certain regions develop while others regress.

16 IMANISHI Kinji, *Shutaisei no shinkaron* (Subjecthood in evolution), Tokyo, Chûô kôron, 1980, p. 198. By the way, the question which in Taylor elicited the hypothesis of the baby sling, elicits in Imanishi the less technical but more social hypothesis of the origin of the couple, that the male had to protect the female who, her hands burdened with the infant, could not defend herself.

17 No. 1135 (April 2012) of *Science & vie* even mentions studies that indicate a lowering of our average IQ.

18 Bénédicte de VILLIERS, *Husserl, Leroi-Gourhan et la préhistoire*, Paris, Pétra, 2010, p. 226.

19 *Kankyô no shutaika, shutai no kankyôka* 環境の主体化、主体の環境化, the formula that is present throughout Imanishi's work, but especially in his late book *Subjecthood in Evolution, op. cit.*

20 *Jiko ryôkai* 自己了解.

21 Paris, Albin Michel, 1922. English translation *A Geographical Introduction to History*, in collaboration with Lionel Bataillon, London 1925.

22 *Op. cit.*, p. 412 in the 1949 edition.

23 *Le Monde cahier science et techno*, December 31, 2011, p. 1.
24 *Les origines de la culture*, Paris, Le Pommier, 2008, p. 60.
25 *La fin de l'exception humaine*, Paris, Gallimard, 2007, p. 218.
26 *Jinkô seimeitai* 人工生命体.
27 MOTOKAWA Tatsuo, *Seibutsugakuteki bunmeiron* (Biology of civilisation), Tokyo, Shinchôsha, p. 207.
28 Paris, Pétra, 2009.
29 Note of translator: in French *être dans la lune*, "to be in (on) the moon".
30 As I mentioned in my review of that profound work, Être humain, être dans la lune? (To be human, does it mean to daydream?) *EspacesTemps.net*, September 20, 2010, link: http://espacestemps.net/document8278.html
31 *Essais et conférences*, Paris, Gallimard, p. 183; translated by André Préau (the passage is taken from *Bauen wohnen denken*). The phrase "as the Greeks observed" forgets about the notion of *topos*, similar to what Heidegger will denounce as *Stelle*.
32 For this theme, see François JULLIEN, *La Grande Image n'a pas de forme, ou du non-objet par la peinture*, Paris, Seuil, 2003.
33 Particularly in *Empire*, Paris, Exils, 2000, *passim*.
34 *Le Monde*, June 3, 2008, p. 17.
35 Christian LAVAL, *L'homme économique. Essai sur les racines du néo-libéralisme*, Paris, Gallimard, 2007.
36 Conclusion reached for the fist time by Antoine PICON, *La ville territoire des cyborgs*, Besançon, Éditions de l'Imprimeur, 1998, however without ontological questioning. This questioning is the subject of my *Histoire de l'habitat idéal, op. cit.*
37 Richard DAWKINS, *The Extended Phenotype. The Long Reach of the Gene*, Oxford, Oxford University Press, 1982.
38 Concept introduced in an earlier book, *The Selfish Gene*, Oxford, Oxford University Press, 1976.
39 Heiner MÜHLMANN, *La nature des cultures. Essai d'une théorie génétique des cultures* (The nature of cultures. Essay on a genetic theory of culture), Marseille, Parenthèses, 2010. The theory of the CMS has been praised to the skies by Peter Sloterdijk; I criticized it in À l'école des toupayes (Learning from treeshrews), *EspacesTemps*, May 16, 2011, link: www.espacestemps.net/document8845.html
40 Edgar MORIN, *Science avec conscience* (Science with a conscience), Paris, Seuil, 1990, p. 100.
41 Stiegler here refers to Gilbert SIMONDON, *L'individuation psychique et collective*, Paris, Aubier, 1989.
42 Bernard STIEGLER and ARS INDUSTRIALIS, *Réenchanter le monde. La valeur esprit contre le populisme industriel*, Paris, Flammarion, 2006, pp. 41 and 44.
43 The German word *Entfernung* means "distance". In *Being and Time*, Heidegger writes *Ent-fernung*, mostly translated as "distancing" or "dis-stancing." For the process I am trying to express, i.e. a to-and-fro, it would be better to use *Ent-Fernung* "distancing–dis-stancing".
44 Note of translator: *rêve* (dream) and *rove*, both of obscure origin, may derive from a common root.
45 *Oneiropoloumen blepontes*, "we dream as we see it", *Timaeus*, 52b3; this is even more the case since *rêver* (to dream) originally meant *to rove*; hence the name of your Land Rover, like that of Curiosity, the Martian vehicle that NASA calls a *rover* even though, as a robot controlled by the System, it can only rove if the system fails. See also below, § 32.
46 寄物陳思.

47 Kisetsukan 季節感. On this theme, KARAKI Junzô, *Nihonjin no kokoro no rekishi* (The history of Japanese sensibility), Tokyo, Chikuma, 1976, 2 vols.; I relied heavily on this in work in *Le sauvage et l'artifice. Les Japonais devant la nature, op. cit.*
48 Note of translator: Lamartine's poem.
49 Note of translator: in French "soul" is *âme*, from the Latin *anima*.
50 Just as we say Leonardo or Raphael, often the first name or pen name is used for men of letters (Norinaga, Bashô …), rather than their family name (Motoori, Matsuo …).
51 Louis DUMONT, Identités collectives et idéologie universaliste. Leur identité de fait, *Critique*, 41 (456), May 1985, pp. 506–518. Why "factual identity"? Briefly, because it is universal to think one is singular, as it is singular to think one is universal.
52 *Op. cit.*, vol. II, p. 170. For Karaki, Norinaga remains a scholar (*gakusha*) in spite of all his efforts, and a bad poet who never reached the emotional power of things he spent his life studying.
53 This ancient transcription is purely phonetic.
54 怜 (*satoi*) means "clairvoyant". It also reads as *awaremu*, to feel *aware*. This sinograph is preceded here by another, very rare one (not in my software), where the heart key is associated with 可.
55 Today this word means bizarre, but in antiquity it meant pleasant, piquant, charming, interesting. According to Jacqueline PIGEOT, *Michiyuki-bun. Poétique de l'itinéraire dans la littérature du Japon ancien*, Paris, Maisonneuve et Larose, 1982, p. 194:

> the words *aware* and *okashi* refer the first to the intensity of the felt emotion and the second to its unexpected character, but do not indicate by themselves the nature of the impression in question. … [*Aware*] designates, *grosso modo*, the emotion provoked in the human heart by the spectacle of the fragile beauty of beings.

56 *Op.cit.*, vol II, p. 156.
57 *Ibid.*
58 "Leaf" in the sense of "tree leaf". The title of the anthology *Man.yô.shû*, transcribed 萬葉集, literally means "anthology of the thousand leaves". Is it a premonition of the relation between poetic language and natural phenomena?
59 Horace, *Odes*, Book III, 30.
60 *Fûdo*, *op. cit.*, p. 36.
61 *Shi e no sonzai* 死への存在, translation of the Heideggerian *Sein zum Tode* (Being towards death).
62 *Sei e no sonzai* 生への存在.
63 *Op. cit.* p. 50.
64 Interview in *Sciences et avenir*, November 2007, p. 107.
65 Ernst KANTOROWICZ, *Les deux corps du roi. Essai sur la théologie politique au Moyen-Âge*, (The king's two bodies. Essay on political theology in the Middle Ages), Paris, Gallimard, 1989 (1957).
66 Alain de Lille, *op. cit.* in § 16, p. 102. On the same page, note 2 explains: "With her two hammers and on the corresponding anvils, Venus must give form to human and animal species and insure their continuity against death [the Parcae]."
67 Alain SUPIOT, *Grandeur et misère de l'État social* (Grandeur and misery of the social state), Paris, Collège de France / Fayard, 2013, p. 29.
68 "… being many", means that, unlike the members of a natural body, the members of a mystical body each have their own personality, as will be emphasized in the encyclical *Mystici corporis* of Pius XII.

69 Paul OURLIAC and Jehan de MALAFOSSE, *Histoire du droit privé*, vol. III, *Le droit familial*, Paris, PUF, Thémis, 1968, p. 418.

70 From this point of view, the indestructible unity of moral body/glorious body may be considered as the symbolization of the mediance animal body/medial body. Cf. St. Paul, Epistle to the Philippians, 3: 19–21:

> Whose end is destruction, whose God is their belly, and whose glory is in their shame, who mind earthly things. For our conversation is in heaven; from whence also we look for the Savior, the Lord Jesus Christ. Who shall change our vile body, that it may be fashioned like unto his glorious body, according to the working whereby he is able even to subdue all things unto himself.

In Christianity, eternal life is not only of the soul but also of the transfigured body. Paul writes above (1: 21): "Christ will be glorified in my body, whether I live or die." While mesology has nothing to say about the divine absolute, the human idea that Christ is life at least symbolizes our mediance.

71 As we have seen in § 20, Watsuji defines ethics as "the study of human interrelation" (*ningen no gaku*). For mesology, this interaction extends to the things in our milieu, which makes it possible to found an ecumenical ethics. This is what I oppose to the ontological aporias of environmental ethics, in *Être humains sur la Terre. Principes d'éthique de l'écoumène* (Being humans on the Earth. Principles of ecumenal ethics), Paris, Gallimard, 1996.

72 Richard DAWKINS, *The Selfish Gene*, Oxford University Press, 1976.

73 Originally, the Yin and the Yang, corresponded to the two slopes of a valley: north-facing slope (ubac, shaded) and slope in the sun (adret, sunny).

74 Hans JONAS, *Das Prinzip Verantwortung* (The responsibility principle),1979.

75 *Let us be towards death*, in Heidegger's language.

76 P. 1302 in Ernst KANTOROWICZ, *Œuvres* (Works), Paris, Gallimard Quarto 2000.

77 *Ibid.*

78 See Pierre HAYAT, Laïcité, fait religieux et société (Secularity, religions and society), *Archives de sciences sociales des religions*, Éditions de l'EHESS, 137, 2007, pp. 9–20 (online).

79 In the Outback of Australia, the Tjukurrpa is the "Dreamtime", the time of the origins and the manifestation of all reality. On this topic see Sylvie POIRIER, *Les jardins du nomade. Cosmologie, territoire et personne dans le désert occidental australien*, Münster, LIT Verlag, 1996.

80 WATSUJI Tetsurô, *Rinrigaku* (Ethics), Tokyo, Iwanami bunko, 2007 (1937), vol. I, pp. 52 ff.

81 *Méditations sur l'Évangile*, La Cène, I, 52e journée (Meditations on the Gospel. The Last Supper, I, 52nd day) (cited in the *Dictionnaire culturel de la langue française* of Alain REY, Paris, Dictionnaires Le Robert, 2005, vol. II, entry "Eucharist").

82 Claude CALAME, Identité et sujet de discours. Soi-même comme les autres (Identity and subject of discourse. Oneself like others), *Revue de théologie et de philosophie*, 138, 2006, pp. 343–354, p. 353.

83 Dany-Robert DUFOUR, *On achève bien les hommes. De quelques conséquences actuelles et futures de la mort de Dieu* ("Finishing off" humans. On some present and future consequences of the death of God), Paris, Denoël, 2005.

84 Dufour quotes (pp. 17 and 18) an article by Doctor Émile DEVAUX, The infantilism of the human with respect of anthropoids, *Revue générale des sciences*

pures et appliquées, 32, 1921, and for the theory itself of human neoteny, see Louis BOLK, *Das Problem des Menschwerdung* (Human genesis), 1926.

85 Ludwig FEUERBACH, *L'essence du christianisme* (The essence of Christianity), Paris, Maspéro, 1968 (1841), p. 249 (quoted by Dufour, p. 94).

86 Note of tranlator: written "d'hommestique" in the original, Dufour's untranslatable pun on *d'homme* (of man) and *domestique* (domesticates).

87 Note of translator: in French, "excluded middle" is *tiers exclu*, "excluded third (term)".

6 Growing together

§ 26. Concrete, concreteness, concrescence

The objective of mesology is to construct *an epistemology of concrete things* as opposed to the abstraction of the modern object. This construction nevertheless presupposes concepts starting with "mediance" and "trajection".

– In short, we begin with abstractions that deny they are abstractions?

– It is not that simple. The opposition of abstract and concrete is derived from the modern idea of abstraction which separated words from things and made things into objects. This abstraction *renders discrete*: it separates the continuous flux of reality into discrete quantifiable units. This goes together with the focus on the object (§ 14). For starters, it *de-concretizes*. To be sure, "concrete" has taken on a meaning very close to "material" and hence objective, so that in English for instance this adjective has now become a substantive. However this is not the issue. Originally, *concretus* in Latin was the past participle of *concrescere*, the verb formed with *crescere* (to grow) and *cum* (with, together). *Cum-crescere* really meant "to grow together". This is what things do when we allow them to generate each other in the field of our existence. There is at the same time a link (*cum*) and a growing (*crescere*). We can say that concreteness, or rather *concrescence* since it is a movement, embodies trajectivity between *genesis* and *chôra*, existing and its milieu. Conversely, the modern separation into discrete units embodies the triumph of the *topos* and the identity principle in individual objects, together with the principle of subject/object dualism. The MOT is the implacable adversary of concrescence, which it forecloses as best as it can.

However, a concrete thing is not only an object that is circumscribed and fixed in its *topos*; it is also the process that involves our existence – our *ek-sistence*. There is coalescence (mutual nourishment)[1] between its being and ours. This implies the surpassing of identity by becoming, in other words the mediation of being by not-being (what one is not yet). This rather Hegelian idea is inherent in the concept of trajection, in which the

S substance is mediated by the non-substance P in the S/P reality. Indeed, P *is not* S; it is accidental, unnecessary and non-essential to it. In Greek in Aristotle's *Metaphysics* (4, 2), it is called *sumbebêkos*. Strangely enough, the term comes from the past participle of *sumbainein*, "to walk together, to go with". This directly evokes the *going-with* of things and of our existence in concrete reality. Except where abstraction reigns, we deal with this going-with and not with discrete objects. This is what gives us the medial *holds* on our milieu,[2] which without trajection would forever remain alien to us. Those holds are the very *being-as* of "S as P" that constitutes the reality of things. Like the Gibsonian affordances they are ambivalent: they are at the same time the holds we have on things and those that things offer us. In other words they are both active and passive at the same time (§ 33). We are far from the in-itself or from the neutrality of objects facing the dualist idea of the subject. The holds motivate us and this motivation extends to our entire milieu. "Holds aggregate into motifs if not into regions"[3] in a continuous movement which is but the concrescence of the *genesis* in its *chôra*.

– But how can these holds motivate us?

– By the effects of the trajection that constituted them historically, hypostatizing in things the way we have of grasping them (P), in other words the expression of ourselves, our inclinations and our judgments. This is what Watsuji's concept of *jikohakkensei*[4] means, the "uncovering of the self" in the milieu (§ 17). The self (*jiko*) in question is fully human, concretely charged with affects and habits, even before it is conscious and capable of becoming an *abstract cogito*.

This trajection not only transforms abstract objects (S) into concrete things (S/P); it also transforms the subject that is its own predicate, i.e. the human being (§ 8), bringing it into conformity with its own milieu (S/P).

– Where does this double effect come from?

– Again, it occurs because the trajective entity (producing the P), i.e. a human subject, expresses itself in its way of grasping things, namely S as P. And so, all along the trajective chain (((S/P)/P′)/P″)/P‴ ..., this "itself" evolves as it emits P′ instead of P, P″ instead of P′, etc. This process brings about the concrescence of the subject with its milieu. This chapter will show some of its concrete manifestations in the ecumene; Chapter 10 will further show its effectiveness at the ontological level in the biosphere as well.

– Let us consider this a promise ... What about these examples?

– Let us start with the encounter between two concepts which will clarify our perspective. In *Fûdo*, Watsuji speaks of "self-understanding" *jikoryôkai*:[5]

> Such actual self-understanding *jikoryôkai* does not consist in understanding a "self" (*ware*) as a "subject" (*shukan*) who feels the cold

> or the heat, or as a subject who enjoys the sight of flowers. In those experiences we do not direct our view to the "subject" (*shukan*). When we feel the cold, our body contracts, we cover ourselves and we get closer to the fire. No, an even stronger interest makes us first dress the children and push old people near the fire. Similarly we work in order to afford to buy clothes or charcoal. Charcoal burners make charcoal in the forest and weaving mills make fabric.[6]

While the concept of self-understanding is undoubtedly inspired by the Heideggerian *Verstehen* (understanding), we see immediately it is more concrete, more geographical and more radical. It is indeed the "self" (*jiko*) who discovers himself and understands himself in flowers, clothes or charcoal; this "self" is something vaster and more profound than a "myself". It is the foundation of all possible "selves". It is the milieu (*fûdo*), in its total concreteness: the going-with of all things with each other, which thus gives rise to our existence. This is precisely what the MOT has rejected.

Those existential holds occur at every level, from the unity of the self in a place to the unity of humanity in the ecumene. The intermediary case which made the modern world is the "taking hold of the land" (*Landnahme*) which, according to Carl Schmitt, made it possible for "the new entity, the state, to eliminate the sacred empire and the imperial reign [*das sakrale Reich und das Kaisertum*] of the Middle Ages. It also eliminated the *potestas spiritualis* [spiritual power] of the pope in the matter of people's rights".[7] Without this hold, there is neither the state nor people's rights in today's sense although the perspective of the MOT forecloses it: isn't all this supposed to be a matter of contracts between individuals?

At another level, the reality that is most directly necessary for human existence (after breathing), i.e., food, is never solely material. In other words, it is never just part of the ecology. Ecological, technical and symbolical at the same time, it falls under mesology proper. As Lucien Febvre has emphasized:

> Nowhere do savages ingest food with some kind of eclectic indifference. There are prohibitions, restrictions and taboos everywhere. ... among the Eskimos of the Centre, it is forbidden to eat land game and sea game at the same time; it is forbidden to keep both killed land animals and whale, seal or walrus meat in one's hut.[8]

It shows the error of utilitarianism which believes that these things are determined by digestion. Of course there is digestion, naturally, in the metabolism that makes life out of the dead matter of the animal body (gathered or killed) taken from the environment. However, no less necessary to the life of the medial body, trajection makes it into concrete human food which remains material but is also named, prepared, ritualized and world-ed. In short: S/P, and not just S.

It is the same with habitat. Concretely, it is not enough to shield the animal body from bad weather, because that body is never just animal; it is human, and its mediance needs to be lodged just as much *according* to a milieu as *in* an environment. This is the reason why a house is never only a container, but also what it contains or rather what goes with it concretely, a family. Traces remain in our languages. In French, *maison* has long meant "the family" and can still be used sometimes in the sense of "company". In the Japanese *ie* and the Korean *chip*, the two meanings of "family" still largely overlap. And what about Chinese where an expression like *dajia*,[9] literally "big house," means "everybody"!

Such things are much more than metonymies, or simply a question of words or even thought. They are the very business of being. The suburban home provides the owner with his existence far more than with lodging. And that is why he remains deaf to the pleas of intellectuals who have been telling him for ages that such a habitat is unsustainable.[10]

Let us not even talk about clothing: when we are told that "clothes do not make the man" it is precisely because concretely the man *is* his clothing, in the larger sense of going-with a certain milieu. This is true for all human reality and not only for social reasons: the reasons are both cognitive and physiological.

Since antiquity we have been familiar with techniques of memorization that associate such and such a thing with another (for example the framework of a discourse with that of a house), enabling them to evoke each other. This going-together is the principle of Proust's madeleine, and not just a literary figure of speech:

> Associating the work of the memorization process to an odor, and diffusing this scent during sleep, strengthens the learning process. The hippocampus, the region of the brain where information transits during memorization, seems to be stimulated by scents.[11]

Since the brain (memory in this case) functions better in concrete situations (here going-with "noeme + odor"), it does not develop normally when they are taken away. Experiments on mice[12] have proven that isolation at an early age harms cerebral wiring; those experiments reveal "irreversible defects at the level of myelin which sheathes the nervous fibers".[13]

Even the law is beginning to recognize concreteness and to consider that the reality of a thing cannot be reduced to an object in its individual *topos*. This could overturn fundamental notions of movable or immovable property:

> And so keys, without which the house would be incomplete, or the panels to close up shops without which things would go missing; or the vases used for the divine service, without which the chapel would

> not be a chapel; all these are part of the building. All those entities are there *ad integrandam domum* [*to integrate the house*]. "Intended to be permanently part of the house", each can be qualified as integral to the building.[14]

The author of this very remarkable thesis even wrote the following: "With this idea of a milieu-thing, the border between real estate and furniture should eventually disappear."[15] ... This would indeed be quite trajective!

§ 27. Places, things and words

Towards the end of her life, Aunt Mimi's memory had greatly declined. She repeated herself often, but according to certain rules that seemed to remain steady. Driving her around in Amos or surroundings, her nephew Luke had noticed that going past this or that place, Aunt Mimi always told the same stories. Inevitably. The story went with the place, revealing the concrescence of what had been a life ... On a different level, this reminded me of those legends which in traditional societies are attached to places. *Legenda* are things that must be said aloud (*legere*) in the place involving them. In return, those stories constitute the place, because, concretely, they are inseparable. Bruce Chatwin in *Songlines* provides a wonderfully beautiful illustration. It takes place in the Outback of Australia. An old Aboriginal man is driven to the place where he wants to die, because that was where the child-spirit came to live in his mother, long ago. The Australian earth is woven with *legenda*, stories told about it since the Dreamtime; approaching the spot, the old man begins to tell the myth, but strangely, as fast as possible. The white people driving him finally understand: since the car was going faster than a man on foot, the words matching the places filing past, were also going faster ...

The preceding story obviously seems like twaddle from a modern perspective that does not consider words as things and places even less; not to mention the word "doghouse", does the word "dog" bark? Nevertheless, those are experienced events (I trust Chatwin), and do express the reality (S/P) of place, words and things in the human milieu, i.e., their concrescence, which by its very nature is their meaning. It is true that words fly and increasingly so thanks to our techniques, but that does not change how meanings originate: their growing-togetherness with things and places, in a given history whose thread continues indefinitely.

– Are you saying that representation and presence, distal and proximal, are all the same thing?

– Yes indeed, in a certain sense! Symbolicity does not deal with *extensio* (note of translator: cf. the Cartesian notion of extended space). The S/P reality cannot be reduced to the S substance, whose representation in words would be arbitrary place-holders, located away from S. In fact, it occurs *in a* milieu, and even *as* milieu, rather than *between* distinct MOTs. The insubstantial P is part of that reality from the proximal stage on; at every degree of physical distance, both in space as in time, P continues to generate the real, and goes together with S no matter what the material vector of this going-together may be, from the most carnal sensations to the resonance of a simple word spoken at the extreme antipodes of the place where the initial S is or was physically. At this point I could invoke what happens in quantum entanglement:

> Two particles that interacted in the past, in certain cases seem to constitute an inseparable totality even when they are very far removed from each other. It is therefore not acceptable to consider that they are individually constituted in the totality they form.[16]

This is probably the nature of meaning. In other words we are dealing with history, interaction in the past between words and things, concrescence rather than the arbitrary projection of a certain signifier on a signified, according to some kind of convention.

"Words": this designates language as well as speech. Of course one can be more precise, even while knowing that many languages do not make the same triple distinction as in French.[17] In Japanese *kotoba* may signify all three; *logos* also in Greek. In reality, language remains an abstraction if it is not embodied in a tongue, just as the latter will remain an abstraction when not actualized in speech by an animal body which itself can say nothing unless coupled with a medial body, in this instance the common system of a language.

All this is not simply *words, words, words*: they belong to concrete situations, in a certain milieu and a certain history, outside of which they mean nothing. Metabasism is the grand illusion of late modernity which believed that words – signs – got their meaning solely from the relation between themselves; in other words, language would be self-originating. To be sure, the relation between signs exists and contributes to meaning; but it no more creates meaning by itself than P creates S/P. Not only do words not create things (contrary to magical thinking), but they are not even independent of them (contrary to the modern view). With respect to things (S/P), words are in the position of P: i.e., what is said concerning an S subject, which is always already S/P (a thing). All this is part of the same trajective chain (((S/P)/P')/P'')/P''' …, which is both historical and medial, both less than and beyond predication in the purely verbal sense of logic which confines itself to rendering discrete – fethishizing – a unique aspect of the global process. See especially the metabasism of French theory, following the momentum of a coup by which language imposed itself as the paradigm of what it was detached from, i.e., reality (S/P), even though concretely it is only one of many predicates (those of our senses, our thinking, our actions). This means absolutizing P in the same mystical leap as Nishida's "logic of the predicate", the enantiomer of the "logic of the subject" which since the scientific revolution has sustained mature modernity. As we have seen above (§ 15), mesology does not absolutize either S or P. Meaning in particular cannot be understood in the framework of Aristotelian logic that absolutizes S and which therefore can only project, at some level, the *topos* of a word onto that of a thing. This is in particular the arbitrary projection (*ex nihilo* and *ex abrupto*) of the value of the sign (P) on pure objects (S) which in our linguistics is supposed to produce the meaning of words. According to the structuralist credo:[18]

> Whatever the moment and the circumstances of its appearance may have been, language can only have emerged suddenly. Things cannot have become meaningful gradually ... a passage occurred between one stage where nothing meant anything to a stage where everything had meaning.

In reality (S/P), meaning does not arise either *ex nihilo* or *ex abrupto*; it is born *ex historia rerum* (from the history of things), in the concrescence of substances and signs, i.e., in the trajection of subjects and predicates: (((S/P)/P')/P")/P"' ..., and so on, ever since the world has been world, that is, at least since life has existed.

Mesology rejects this structuralist and post-structuralist doctrine of suddenness, which is just a materialist metaphor for Creation; and of course mesology takes Bourdieu's[19] side when he reproaches Saussure for having fetishized language by abstracting it from the social conditions of its utterance. Saussure thus separated language and discourse and kept only the former as an object for linguistics. An object indeed, planted in its *topos* and removed from its *chôra*. Clearly, we are in the MOT perspective. Even worse: while priding himself on being post-structuralist, Derrida understands the notion of *chôra* itself from the MOT perspective by isolating it in the *topos* of the Platonic text that mentions it. He abstracts it from any going-together in the milieu where at that time, beyond this *topos*, such a term could be used in Athens.[20] In short, Derrida abstracted it, cutting it off from what made it meaningful.

As for the birth of language, mesology can only take a principled stance: it is an additional, and therefore cumulative, trajection of the semiosphere that is inherent in life, as biosemiotics has shown:

> The semiosphere is a sphere just like the atmosphere, the hydrosphere and biosphere. It penetrates all corners of those other spheres, by incorporating all forms of communication: sounds, odors, movements, colors, forms, electric fields, thermal radiation, waves of all kinds, chemical signals, touch, etc. In short, signs of life.[21]

Life is what assembles signs (P) and substances (S) to make them concretely into things (S/P). Starting with this going-with, signs can transmit meaning.

Little by little, evolution and history transformed this transmission from biosemiotics into properly human language, together with hominization. Far from being a sudden and arbitrary projection, signs progressively became words in this trajection, and so consubstantial with things. Words are things because the latter are not just objects: they are holds, those of our existence *with* our milieu.

The MOT shut off this history. As a result, it is incapable of understanding meaning, and can only fall into an aporia if it rejects the mystical leap of divine creation. The quote from Lévi-Strauss mentioned above is clear: in "between one stage where nothing meant anything" and "a stage where everything had meaning" Lévi-Strauss hypothesizes a "passage" that anthropology cannot deal with: "Resulting from a transformation whose study does not belong to the field of the social sciences, but to biology and psychology,[22] a passage occurred."[23] It is very strange that anthropology, the human science par excellence, should remain silent faced with what characterizes humans, the *zôon logon echôn* (the animal that has language), and leaves it to zoology of all things! In fact, since the author was not a creationist, this follows directly from the tacit postulate of a contract between disembodied individual subjects, talking *cogitos* without being alive, dried up in their MOT, deciding arbitrarily one day, that some object would be called this and another that. It looks like a contract fallen from heaven (minus the Eternal) while in fact meaning has risen from the Earth, in about four billion years, in the contingency of evolution and history. Through trajection and not projection!

§ 28. Japanese milieu, discourse and haiku

To conceive of the meaning of words as simply the projection of human sovereignty over things, abstracted as objects, is obviously an expression of modern dualism. Beyond that, it is the expression of the implicit ontological structure which in the major European languages, together with other historical factors, has fed the transcendence of the pronoun "I" with respect to concrete circumstances of utterances (Chapter 1). It is worth wondering what happens in languages and milieus where the speaker's speech functions according to different principles. We began with a comparison with Japanese; let us continue.

Commenting on Benveniste's formula "I is who says I", imitating the *ehyeh ascher ehyeh* (*sum qui sum*) in Exodus (§ 3), Dany-Robert Dufour writes:

> In fact Benveniste attributed to the subject the self-referential or unitary definition previously attributed to the big Subject [God]. … I is who says I "is" a proposition in which the predicate consists in a return to the subject where the I is defined by himself. … By discovering this point of return, Benveniste has discovered the Archimedean

> point of a new linguistics, i.e., of speech, in other words the tipping point at which language was converted into discourse.[24]

Along with Searle, Austin, Ducrot, Labov, Grice and Jakobson, Benveniste thus participates in a current that has re-concretized linguistics by showing that language first has to be a discourse, spoken by a subject in certain circumstances. Jakobson speaks for instance of "shifters" (such as *I, tomorrow, here* ...) that link this discourse to the concrete existence of the speaker. However, languages relate very differently to the need for such shifting. In other words, they are more or less abstract or concrete. Compared to French, Japanese is a very concrete language, to the point where, as we have seen (§ 4), an utterance such as "Hanako is sad" is impossible unless the speaker is Hanako herself. It can only be the singular discourse of a certain person, and not of another since it deals with her interiority. Even though this confirms the principle of Benveniste's "I is who says I", it follows an entirely different ontology than that of the transcendence of "I", by depending on the concrete circumstances, atmosphere and milieu of that discourse. This amounts to a *sum id, ubi sum*: I am where I am, or at least to Ortega y Gasset's *yo soy yo y mi circunstancia* (§ 6).

Japanese culture never ceased to elaborate, develop and update this linguistic disposition towards concreteness in the course of its history. Medieval literature for instance made great use of *makura-kotoba* 枕詞, those "pillow-words" that Jacqueline Pigeot defines as "frozen words or phrases that cover a word, from which they are as inseparable as a 'Homeric epithet'. The original meaning of the phrase may eventually disappear; for instance *kusa-makura*, 'grass pillow' is now used regarding sea voyages".[25]

Even more remarkably, those pillow-words evolved into poem-bearing toponyms *uta-makura* 歌枕, certain "famous places" (*meisho* 名所) that immediately evoke such or such a thing, event or feeling. In these poem-bearers:

> Geography, that is the exterior objective world and mind penetrate and confirm each other, color each other to determine the poem about which one could say it escapes any gratuitousness by its very order. The poet's choice of those sites that have the status of *meisho*, can thus be better explained from this perspective: is it done in order to avoid even more radically the temptation of gratuitousness? The *meisho* is not only fixed geographically, it is also determined by a rhetorical tradition: introducing a famous place name in a poem means assuming this tradition and inserting oneself in a world of images and correspondences.[26]

In many *meisho* the toponymic function (the geographical identification of a place) may even be overshadowed by the evocative function, that is, the evocation of a feeling by reference to some historical or literary precedent, a feeling that in principle (abstractly) cannot be located but which is here taken concretely, literally. Those renowned [*re-nommés*] places, in other words, are more-than-predicated, exceed the limits of their *topos* in the *extensio*: they become the reader's own feelings.

Obviously, in the course of this trajection the *meisho* cannot escape the fate of all predicates: history tends to substantialize them and ossify them into pure conventions. Towards the end of the Middle Ages, Shôtetsu (1381–1459) for instance, could boast about juggling with the *meisho* without even knowing their location in the territory. "Tatta" and "Yoshino" respectively were par excellence the *meisho* of red maple leaves in the fall and cherry blossoms in the spring; Shôtetsu would use such toponyms to evoke these themes, but without any geographical implication: "Who says red leaves says Tatta, who says flowers says Yoshino, and that is enough."[27] Since place had been substantialized in theme, and become pure formalism, Bashô would systematically visit the *mesiho* on foot three centuries later, revitalizing poetry through the Earth itself.

To be sure, such phenomena are not unknown in our language: it proves that concreteness is universally shared. For instance, gymnasts who use magnesium (magnesium carbonate) generally do not know that the word comes from a city in Asia Minor (today Inekbazar) which had abundant magnetic minerals. Even in antiquity, *magnêtis* meant "magnetic". Taking in the image, one could say that a certain magnetism makes territories and their contents cohere, including words and feelings. This metaphor is of course just another name for mediance.

However, various cultures have not accepted such concreteness to the same degree. It was foreclosed by the *cogito* and the MOT. Japan on the other hand, made it into the fundamental rule of the most universally known literary genre, which is still massively practiced by the Japanese themselves, and even in other languages and countries: the haiku. It was codified around the same time that by contrast the abstract paradigm of modernity was established at the other end of Eurasia.[28]

We met a haiku before in § 4. We immediately saw its most remarkable feature: it is a short poem, consisting of three verses of five, seven and five feet. As a rule, it "sings the phenomena of the natural world produced by the seasonal cycles, as well as the phenomena derived from them in the human world".[29] In other words it sings of the Japanese world. A vast topic! History put some order into this, in particular with the works of Ihara Saikaku (1642–1693) and Matsuo Bashô (1644–1694), continuing up to Masaoka Shiki's (1867–1902) work. In the 20th century the genre acquires its present form, in spite of the violent postwar controversies in which the critic Kuwabara Takeo decried its conformism. The fact remains that in the 21st century, the Japanese continue to greatly cherish it.

A short mass circulation guide defines it as follows:[30]

> The haiku is a poem with a fixed form (*teikeishi* 定型詩), consisting of five–seven–five feet and has as its rule (*kimari* 決まり) the inclusion of

> a seasonal word (*kigo* 季語). Seasonal words are words that express the season, and are inventoried in books of a genre called *saijiki* (歳時記, "annals of years and moments", almanacs, guides of seasonal words).

Remember for instance the seasonal word *fûrin*, the wind chime in the haiku of § 4. Seasonal guides began to appear in the 17th century. According to André Delteil, the most voluminous one today has around seven thousand entries. One of the best sellers is the pocket dictionary of Ôno Rinka[31] with close to three thousand entries, each comprising a definition of several lines, and six or seven examples of haikus taken from the literary heritage to illustrate a particular seasonal word and to give advice on how to better appreciate the poems. The introduction tells us:

> In the last few years, we have seen a boom in these guides. Almost every year, new ones appear that claim to have four to five thousand seasonal words and pride themselves on including so many. Each necessarily adds new terms, which reflects the evolution of our lifestyle. For example the mountain climbing fashion leads to the search for terms for mountain flora, names of herbs or wild birds, which also leads to the search for customs related to these phenomena that are part of our daily life. This movement cannot be stopped. There are good reasons for new seasonal words to appear, and the compositions that include them should be recognized.[32]

It is clearly a living custom that evolves at the same pace as Japanese everyday life. The guides contain more than traditional motifs like the wind chime. For example, among the seasonal words listed in the "Winter" volume of the *New Large Seasonal Guide of Japan*,[33] you can find on p. 137 the snowmobile, called *setsujôsha* or *sunômôbiru* in Japanese, both recognized as seasonal words, illustrated by the following haiku by Wakaki Ichirô:

雪上車	*Setsujôsha*	Snowmobile
丘のうねりの	*oka no uneri no*	From hill to valley
なりに馳す	*nari ni hasu*	It zips along

The haiku continually takes in new things, but according to strict rules; the basic point of reference remains the natural framework of the Japanese milieu, according to the seasonal progress of the most diverse scenes in life. This literary genre illustrates above all the concrescence of moments, things and events in a particular milieu which it puts in order (which it worlds) by staging them. The seasonal guides thus become marvelous handbooks for learning about nature, and the ancient and new customs that go with it. These are grammars of concreteness, choreographies of everything that constitutes a milieu. They organize the going-together-with well beyond words; the seasonal guides collect real *mesonomes*, codes of mediance that just as the *Grevisse* [note of translator: French grammar, the ultimate reference book] collects accepted practices in the French language.

This is really a syntax that goes beyond language alone. The haiku and seasonal guides are thus part of a general tendency of Japanese culture, namely to organize the non-verbal optimally since the verbal as such is less honored, compared with the Greco-Latin tradition where words are abstracted from the concrete milieu, as language rather than discourse. This is precisely why the haiku can be brief; since the milieu is a given, the verbal does not need to be verbose. The implicit is enough.[34]

Since the dawn of its history, hasn't Japan defined itself as the country blessed by the gods where it is not necessary to elevate words (*kotoage senu kuni*)?[35] The norm here is to deal with behavior, circumstances and milieu; hence the development of numerous syntaxes that are extra-linguistic. They touch on all areas of social life, but are particularly elaborate in the so-called traditional arts, like the floral arts, and the martial arts such as karate. They are called *kata* 型, a word whose basic meaning is the general or potential form of particular (*sugata* 姿) or effective forms (*katachi* 形). Those collective forms channel individual ways of doing things. They are matrices, both temporal (particularly in karate for instance where they determine the sequence of gestures) and spatial (particularly in the floral arts where they determine the topologies between elements). As in speech, the individual spirit and common syntax meet: the *kata*, makes possible, sustains and orients personal expression which nourishes the *kata* in turn. The creation of a work may or may not be sparked through conflict, the structural moment of the individual and collective dimensions of the *kata*. This is especially the case in the haiku and its social organization today that according to Delteil provides a framework for

> several million people, among whom a hard core gathers around some eight hundred specialized magazines. The *haijin*, those who practice the haiku, get together at least once a month to vote on the works presented by their colleagues; the rule forbids authors to choose verse composed by themselves; only at the last moment is the authorship of each of the poems revealed. The selection process makes the authors into readers, which is a phenomenon belonging to a group art (*za*) where resonances are in play that often go beyond the understanding of the newcomer.[36]

This is how in a manner quite the opposite of the MOT, Japanese mediance is cultivated, which tends to make Japanese into what they are, i.e., where they are: Japan. *Id fieri, ubi sunt* (to become that where they are) … It is that ideal of concrescence that the Nishidian "logic of place" once tried to grasp (§ 11).

§ 29. The co-creation of things

A world where things go together is obviously not a world where one thing causes the other. It is a phenomenal world rather than a physical world; does that mean it is merely a premodern world? In some respects, yes; but from a mesological point of view, we need to look further. Husserl writes that "the primordial Earth does not move",[37] (*Die Ur-Arche Erde bewegt sich nicht*), an expression that appears in a manuscript of May 7–9, 1934, entitled "Reversal of the Copernican doctrine in the interpretation of the usual view of the world. The primordial Earth does not move. Fundamental research on the phenomenological origin of corporeity, of spatiality in the original sense of the natural sciences". Here he explicitly takes the opposing view of Galileo's legendary protest *Eppur, si muove!* (And yet, it moves!) that he supposedly mumbled after having been forced to recant his heliocentric theory at his trial before the Inquisition in 1633. Obviously Husserl does not intend to go back to the geocentric view preceding the Copernican revolution, or more widely deny the contribution of the scientific revolution. The reversal he talks about is not the negation of the objective universe but a re-centering of the lived world (*Lebenswelt*); this amounts to distinguishing it from the physical universe that the scientific revolution had, at least in principle, radically de-centered by making it into a pure object. This reversal then is a recovery – like *u Riacquistu*, as those on the banks of the "Tavignanu" call it[38] – but it does not abolish the de-centering of the modern scientific vision. He outlines its limits, namely its fundamental incapacity to take into account the objective focus of the phenomenality of concrete existence. It is precisely Husserl's intention to effect such a phenomenological re-centering.

This is quite a revolution in the relation between the human being and nature, compared to the ontological dualism of modernity, formulated by Descartes with the distinction between *res cogitans* and *res extensa*, in other words, the dichotomy between the subjective and the objective. To this abstract and reductionist opposition, phenomenology intends to substitute the study of "the actual way in which every object is constituted for our view. In that sense it aims at expressing the real through the links that allow us to know it".[39] For mesology these are the trajective chains that in our eyes establish nature (the environment, the *Umgebung*) as a theme of consciousness (the milieu, the *Umwelt*).

This is more than just one more philosophical position. We can indeed speak of a "phenomenological revolution" in as far as Husserl's work is contemporary with, and not independent of a profound change in the queen of the natural sciences, physics itself. This change amounts to surpassing the modern classical paradigm of Kepler, Galileo and Newton, both at the level of the infinitely large with Einsteinian cosmology, and the infinitely small, with quantum mechanics. One of the protagonists of this transcending process, Werner Heisenberg (Nobel prize in physics,

1933), defined its essence in terms that directly evoke the phenomenological principle:

> If it is possible to speak of the image of nature according to the exact sciences of our time, it must be understood as the image of our relations with nature rather than the image of nature. … It is above all the whole network of relations between man and nature that is the object of this science. … As science ceases to be the spectator of nature, it recognizes its own participation in the reciprocal actions between man and nature. The scientific method that chooses, explains, and puts in order, admits the limits imposed by the fact that the use of the method transforms its object and that as a consequence, the method cannot be separated from its object.[40]

To say that the method cannot be separated from the object: doesn't this mean that we are dealing here with a going-together-with, a growing-together that breaks with dualism? In as far as modernity is largely defined by the effects of the scientific revolution and its dualistic paradigm, the recognition of concreteness signifies a real "overcoming of modernity". Even so, can we compare this surpassing with that claimed by the Kyoto School under the name of *kindai no chôkoku* 近代の超克? Only to a certain degree. For mesology, the logic of place (*basho no ronri* 場所の論理), or "logic of the predicate" (述語の論理) evoked by that School (§ 11) only ends up throwing subject-centered Aristotelian logic into its enantiomer. The trajection that produces concrete reality amounts to the synthesis and even to the sublation (*Aufhebung*) of both logics.

We will deal with these questions more directly in Chapter 7; let us here however focus on this powerful image "the Earth does not move" and compare it with "the use of the method transforms its object". The latter amounts to saying that the object no longer exists in itself, in that famous *an sich* that Kant distinguished from the phenomenon, but exists precisely *as* the phenomenon itself. To be sure, it is not just any phenomenon because it presupposes the rigorous protocols of the scientific method; however, in its principle, this *as* is strictly homologous to trajection which institutes S *as* P in the ecumene. It is the same trajection that, from one star among billions of others and prior to the radical abstraction of the Copernican reversal, finds concretely *the* sun rising at *the* Earth's horizon.

– In this sunrise, where is the method?

– In the objective functioning of the trajection that made it possible for this fact to appear to us as it does now, after four billion years of evolving adjustments, in short, in this landscape and this historicity.

– Isn't your argument a bit specious? Who would be the subject, the agent applying that method?

– Nature, of course! but on a *scale of subjectity* and correlatively a space-time scale which is not ours for us as *cogitos*. I will try to develop this strange idea in Part III. In the meantime, let us return to the sunrise. For physics it is the effect of a relative movement. I move with the Earth, so for me the Earth doesn't move, and it is the sun that seems to rise. Elementary, my dear Watson! The problem with this abstract principle, is that we will never get at the concrete reality of dawn, nor that of *Morgen im Riesengebirge* (Morning in the Giant Mountains).[41] Never! Not only is the going-with of all these elements of such a scene far too complex to deduce this concrete reality but also because this going-with is not a matter of deductive or causal inference. In the scene of a dawn, there is far too much contingency, starting with the fact that I am there when I could be elsewhere or there at a different moment. See if you can quantify what follows.

> I remember having spent a delightful night outside the city, on a road along the Rhône or the Saône, I don't recall which of the two. Terraced gardens lined the road on the other side. As it had been very hot that day, the evening was charming; dew moistened the wilting grass; there was no wind in that quiet night; the air was fresh without being cold; the sunset had left a reddish haze whose reflection made the water look pink. The trees on the terraces were filled with nightingales calling to one another. I was walking in a kind of ecstasy, surrendering my senses and my heart to these pleasures, while sighing with some regret that I had to enjoy them alone. Absorbed in my sweet dreams, I continued walking well into the night without realizing I was tired until I finally noticed it. Voluptuously I laid down against some kind of a recess or sham door sunk into a terrace wall. The sky of my bed was shaped by the tree, with a nightingale just above my head; its song lulled me to sleep. My sleep was sweet and my awakening sweeter still. It was broad daylight; I opened my eyes and saw water, greenery and an admirable landscape. I got up, shook myself and as I felt hungry I cheerfully walked towards town. I had decided I would use my remaining pennies to buy a good breakfast. I was in such a good mood that I sang all along the way.[42]

Those are just words. Far from a real sunrise! Nevertheless, you can see how everything here, from the river stream to the song of the nightingale, without forgetting the Earth's rotation or Jean-Jacques' heart, corresponds to *one* single experience; thoughts, images, multiple sensations, the coenesthesias and the continual synesthesias create a whole atmosphere. Where are the causes? Nowhere, because they are everywhere. In this concrete harmony, which is no other than reality, each thing becomes the *condition* of the others and all respond to each other at the same time. They *co-originate* each other.

– "Co-originate!!" where are you taking us?

– Don't worry! To a territory marked out long ago, which we will reach by building on the Copernican revolution!

Notes

1 From *cum*, with, mutually, and *ali*, to nourish oneself.
2 I developed the theme of medial holds in the ecumene in *Écoumène, op. cit.* (see above, Propos), chap. 6 ("Prises") (Holds).
3 Baptiste de COULOMB, *Médiance, pouvoir et agriculture de proximité. L'exemple des AMAP (Associations de maintien de l'agriculture paysanne franciliennes)* (Mediance, power and local agriculture. The example of AMAP (Associations of Maintenance of Farming Agriculture in the Île de France), MA in Comparative Studies of Development), EHESS, September 2006, p. 29. "Region" here is an allusion to the Heideggerian *Gegend*, but understood in its geographical reality.
4 自己発見性.
5 自己了解.
6 *Fûdo, op. cit.*, pp. 24–25.
7 Carl SCHMITT, *Le nomos de la terre. Le droit des gens du jus publicum europaeum* (The nomos of the Earth in the international law of the jus publicum europaeum), Paris, PUF, 2001 (*Der Nomos der Erde im Völkerrecht des* Jus publicum europaeum, 1950), p. 127.
8 Lucien FEBVRE, *La Terre et l'évolution humaine. Introduction géographique à l'histoire* (Earth and human evolution. Geographical introduction to history), Paris, Albin Michel, 1949 (1922), p. 206.
9 大家. In Japanese, *ie* is written as 家 (read *jia* in Chinese). By the way, this sinograph consists of the two elements "roof" (above) and "pig" (below), and so literally means "pigsty"!
10 See the publications of the program "Unsustainability in human settlements" (2001–2010): A. BERQUE, Ph. BONNIN, C. GHORRA-GOBIN (eds.) *La Ville insoutenable* (The unsustainable city), Paris, Belin, 2006, A. BERQUE and S. SUZUKI (eds.) *Nihon no sumai ni okeru fûdosei to jizokusei* 日本の住まいに於ける風土性と持続性 (Mediance and sustainability in Japanese settlements), Kyôto, Nichibunken, 2007; A. BERQUE, Ph. BONNIN, A. de BIASE (eds.) *L'habiter dans sa poétique première* (Settling in its prime poetics), Paris, Donner lieu, 2008; A. BERQUE, N. FROGNEUX, B. STADELMANN, S. SUZUKI (eds.) *Être vers la vie. Ontologie, biologie, éthique de l'existence humaine* (Being towards life. Ontology, biology, ethics of human existence) Tokyo, Maison franco-japonaise (*Ebisu*, 40–41), 2009; A. BERQUE, Ph. BONNIN, A. de BIASE (eds.) *Donner lieu au monde. La poétique de l'habiter* (Giving rise to the world. Poetics of settlement) Paris, Donner lieu, 2012.
11 *Science & vie*, 1076, May 2007, p. 36.
12 Conducted by a Harvard team and published in *Science & vie*, 1142, November 2012, p. 38.
13 *Ibid.*
14 Sarah VANUXEM, *Les choses saisies par la propriété* (Things seized by ownership), Paris, IRJS éditions, 2012, p. 239.
15 *Op. cit.*, p. 243.
16 Michel SERRES and Nayla FAROUKI (eds.) *Le Trésor. Dictionnaire des sciences* (The treasure. Dictionary of sciences), Paris, Flammarion, 1997, p. 642. This inseparability was proven experimentally by Alain Aspect in 1983 on a pair of photons.
17 Note of translator: the triple distinction in French is as follows: *langue* is language as a system; *langage* is a particular form of language use; *parole* is speech.
18 From the pen of Claude LÉVI-STRAUSS, Introduction à l'œuvre de Marcel Mauss (Introduction to the work of Marcel Mauss), in Marcel MAUSS, *Sociologie et anthropologie* (Sociology and anthropology) Paris, PUF, 1950, p. xlvii. Half a century later, this passage is still cited uncritically by Sylvain AUROUX,

La Philosophie du langage (Philosophy of language), Paris, PUF, 1996, p. 45. More recently, the same thesis has been stated by Richard KLEIN and Blake EDGAR, *The Dawn of Human Culture*, New York, John Wiley & Sons, 2003, for whom language suddenly appeared around 50,000 years ago, with a parallel expansion of symbolic function, after the mutation of the gene Fox P2. From the same genetic perspective, this date has been pushed back 1.8 to 1.9 million years by Karl DILLER and Rebecca CANN in Rudolf T. BOTHA and Chris KNIGHT (eds.), *The Cradle of Language*, Oxford University Press, 2009.

19 Pierre BOURDIEU, *Ce que parler veut dire. L'économie des échanges linguistiques* (What speaking means. The economy of linguistic exchanges), Paris, Fayard, 1982.

20 Jacques DERRIDA, *Khôra*, Paris, Galilée, 1994. I developed my critique, The *chôra* in Plato, pp. 13–27 in Thierry PAQUOT and Chris YOUNÈS (eds.) *Espace et lieu dans la pensée occidentale* (Space and place in western thought) Paris, La Découverte, 2012. See also below, § 32.

21 Jesper HOFFMEYER, *Signs of Meaning in the Universe*, Bloomington, Indiana University Press, 1996 (1993), p. vii.

22 In the sense of course of physical psychology, resembling what we would call cognitive science today.

23 LÉVI-STRAUSS, *ibid.*

24 DUFOUR, *op. cit.* in § 25, p. 263 and 264.

25 PIGEOT, *op. cit.* in § 24, p. 26, note 31.

26 *Op. cit.*, p. 111.

27 *Tada momiji to ieba Tatta, hana to ieba Yoshino to ieba tariru* ただ紅葉といへば龍田、花といへば吉野といへば足りる. Cited in KARAKI, *op. cit.* au § 23, vol. I, p. 85.

28 For more general information on the haiku (the term became generalized only during the Meiji period, by contracting *haikai-hokku*), see André DELTEIL, Haikai, haiku, pp. 229–231 in A. BERQUE (ed.) *Dictionnaire de la civilisation japonaise* (Dictionary of Japanese civilization) Paris, Hazan, 1994.

29 DELTEIL, *art. cit.* p. 231.

30 TOKITA Tomoya, *Rokujussai kara no tanoshii haiku nyûmon* (Introduction to the pleasures of haiku after sixty), Tokyo, Jitsugyô no Nihon-sha, 2008, p. 20.

31 ÔNO Rinka (ed.), *Nyûmon saijiki* (Introductory guide of seasonal words), Tokyo, Kadokawa shoten, 23rd ed., 2004 (1980).

32 *Op. cit.*, p. 1.

33 IIDA Ryûta et al. (eds.) *Shin Nihon dai saijiki* (New Year, Spring, Summer, Fall and Winter), Tokyo, Kôdansha, 5 vols.

34 I agree here with Jacqueline PIGEOT, *Questions de poétique japonaise* (Questions of Japanese poetics), Paris, PUF, 1997, p. 7: "It is now easier to understand the vitality and success of the 'short poem'. … If the formal 'units' of Japanese literature (like the *waka*) are short, it is not because of some kind of interest in the miniature, but because these units do not have the independence of units in western literature (ballad, sonnet, etc.); they are interdependent and are defined by their relation to other elements: other poems or fiction." As we will see those "other elements" go well beyond literature alone; they deal with a whole milieu.

35 *Man.yôshû*, song 3253–3254, attributed to Kakinomoto no Hitomaro (active at the beginning of the 7th century). The first sentences say: *Ashihara no mizuho no kuni wa kamu nagara kotoage senu kuni* 葦原の瑞穂の国は神ながら言挙げせぬ国, literally "The country of reed beds with young rice stalks is the country where thanks to the gods, words are not elevated." This is generally understood as: Japan is so favored by the gods that it is not necessary to pray to them.

36 DELTEIL, *art. cit.*, p. 231.
37 Edmund HUSSERL, *The Primordial Earth does Not Move* (*Die Ur-Arche Erde bewegt sich nicht*, 1940) Paris, Minuit, 1989.
38 The Tavignano (Tavignanu in Corsican) flows in the heart of Corsica, as the Genoese would say.
39 Gérard WORMSER, entry "Phénoménologie" (Phenomenology) in Michel BLAY (ed.) *Dictionnaire des concepts philosophiques* (Dictionary of philosophical concepts), Paris, Larousse/CNRS éditions, 2006, p. 615.
40 Werner HEISENBERG, *La nature dans la physique contemporaine* (*Das Naturbild der heutigen Physik*, 1955) Paris, Gallimard, 1962, pp. 33–34.
41 This painting from 1810 or 1811 by Caspar David Friedrich is today at the Alte Nationalgalerie in Berlin.
42 Excerpt from Book IV of the *Confessions* of Jean-Jacques Rousseau, who speaks here of his last long hike from Paris to Chambéry in 1731, when he was 19 years old.

7 Including the middle

§ 30. The return of an old book

This book is a child of contingency, which I believe could be productive or even poetic. It would have been a different book without an encounter with another book I bought in Sendai shortly after its publication in the fall of 1974, *Rogosu to renma* (*Logos and Lemma*), by Yamauchi Tokuryû.[1] I stopped reading it after a few pages. Even though its theme had interested me, I was too green intellectually at that time to understand it, or to discern its importance. I was working at the time on my thesis on the geography of the Hokkaidô countryside, dealing with rice fields and ice floes rather than with the tetralemma and Nāgārjuna.

I remained too green for *Logos and Lemma* for many years. Although I was sure that the book carried some meaning for me, I postponed reading it and eventually lost it in the course of the numerous moves that offered opportunities to abandon part of my always excessively large library. However, like lizard tails, libraries grow back. And so, 36 years later, what shows up in the mail at my home in Palaisseau but this book, now unobtainable in Japan, sent by the landscape architect Nakamura Yoshio … He had found it in Kanda in a used bookstore and, browsing through it, was fascinated. For some unfathomable reason, he thought that I was the one who would make good use of it.

It was the same book, in perfect condition apart from the fact that the page edges had darkened. The gilding was still superb. The book had actually remained so intact that inside, forgotten amidst the pages, there were two traces of its former owners' life. First a postcard, with a stamp dated in Kyoto, Sakyô-ku, on October 16, 1979. It was an invitation to the addressee to two conferences for the following November 2 organized by the Kyoto Philosophy Society, one on "the two processes of memory", and the other on "the Japanese state". Also included was a leaflet announcing a seminar from July 29 to August 2, year not specified, in a spa resort called Shin Hirayu deep in the Hida mountains (the "Japanese Alps"). If the postcard stamp was any indication, Yamauchi, who lived from 1890 to 1982, must have been 89 years old. The bottom half of the leaflet, a reply-coupon, had been cut off. The upper half said:

Announcement of a summer seminar on the work of the philosopher Yamauchi Tokuryû
Honorary professor at the University of Kyoto
Organized by
The philosophical research group
Of the Waseda University,
Building 3, main area

> Dr. Yamauchi will honor with his presence the seminar organized on his work, for people interested in philosophy, on a superb site deep in the Hida mountains, the Alps of the North. The presentations will not only focus on Nishida's philosophy, its reception and repercussions, but also on further developing it systematically, linking it closely to "from the thing to the fact, the problem of causality".[2] Through his research in the history of philosophy, Professor Yamauchi has been a pioneer in numerous, wide-ranging fields with his work on comparative Eastern and Western philosophy, and has also opened up an original philosophical field with his theory of the object. He recently published a book that has had a great impact, *Logos and Lemma*. People interested in philosophy are cordially invited to this seminar, where we will meet with him daily.
>
> Books by Professor Yamauchi:
>
> *Genshôgaku josetsu* (*Introduction to Phenomenology*), *Sonzai no genshô keitai* (*Forms of the Phenomenon of Being*), *Taikei to tensô* (*System and Implementation*), *Girisha no tetsugaku* (*Greek Philosophy*, 5 volumes),[3] *Jitsuzon to shoyû* (*Existence and Possession*), *Rogosu to renma* (*Logos and Lemma*), and many others. [added were details of the material organization of the seminar]

It is strange that in a human life, the same opportunity should present itself twice, more than a generation apart. If instead of going to the vast land of Hokkaidô, I had come to Japan to write a thesis on Nishidan philosophy, perhaps one thing leading to another and even to the depth of the Hida mountains, I would have attended this seminar on causality with Yamauchi Tokuryû, because he was one of the official disciples of the intellectual leader of the Kyoto School. However, I was a geographer and his name was unknown to me. What started the connection was just the title of the book *Logos and Lemma*, discovered on a fall day on the shelves of the cooperative library of the University of Tôhoku, whose science faculty had invited me.

If I had read the book at that time, I would undoubtedly have investigated sooner a word of which I only knew the popular usage: *engi* in expressions such as *engi ga ii* (*warui*), "boding well (or ill) for the future", or *engi wo katsugu*, literally "to carry *engi*", that is, "to be superstitious". Nothing there to tempt my positivism! As we will see in this chapter, this word *engi*, the Japanese pronunciation of the Chinese *yuanqi* 縁起, is in fact one of the conceptual pillars of the Buddhism of the Great Vehicle, where it translates the Sanskrit *pratītya samutpāda*. This has been rendered in our language by expressions as varied as: conditioned co-production, consecutive production, interdependence, production by conditions, production in mutual relation, dependent event, production in convergence, going in function of, dependent origination, conditional genesis, dependent co-advent, being by co-emergence ... or even by *contingency*, the word with which this chapter opened and which we will often see.

There is perhaps some *pratītya samutpāda*, some unfathomable reason why a book, whose importance I had sensed but not grasped in my youth, and ignored for three decades to the point of losing its trace, should come back to me in the mail, at an age when I would finally be able to benefit from it. Yes, it showed the contingency of "what can be or not be". Between chance and necessity, between the mechanism of causes and the vagaries of the imaginary, there is, in what causes the things of our world to exist, a middle ground that western thought has neglected, but which nevertheless constitutes the fabric of human reality, and even of the realities of the biosphere in general. This will be my thesis in Part III of my book.

The present chapter will explore that domain of relations in the human world. While the discourse is mesological, it will include a clarifying dialogue, based on systematic references to Yamauchi's book, which I consider one of the great texts of the 20th century.[4] History must take the blame for the fact that he is almost unknown outside Japan, and little known even in Japan. Yamauchi has been linked to the Kyoto School. He was Nishida's student, and after the war he became the first chair of the philosophy department where Nishida had taught, even though Yamauchi's thought is not based on Nishida's. And so he did not benefit from the exceptional revival that Nishida and his heirs have known since the 1980s, after their banishment for having compromised with ultra-nationalism. In a way, Yamauchi remained ostracized even though his thought had nothing to do with "overcoming modernity". On the contrary, he tried to effect a synthesis between the western mode of thinking, which he saw in the logos, and the oriental mode of thinking, which he saw in the lemma. In other words, while overcoming modernity was seen as a reaction aimed at overthrowing western intellectual supremacy, Yamauchi wanted to give the lie to Kipling's famous saying "Oh, East is East, and West is West, and never the twain shall meet, / Till Earth and Sky stand presently at God's great Judgment Seat".[5]

It was in the same spirit, 30 years after the seminar on "From the thing to the fact, the problem of causality", that with the help of the Maison Franco-Japonaise, led then by Marc Hubert, I organized a seminar in memory of Yamauchi in the same inn of the same resort, with almost the same theme: "From the thing to the fact, the question of the milieu."[6] It was impossible to hold this meeting at exactly the same dates, but instead, symbolically, we chose July 7, 2011. Why July 7? Because the seventh night of the seventh month is the celebration of Tanabata 七夕, whose *Celebrations and Rites of the Four Seasons in Japan*[7] reminds us of the following story:

> According to a legend of Chinese origin, only on the seventh night of the seventh month can the two stars of the Weaver (Vega of the Swan constellation) and the Herdsman (Altair of the Eagle constellation) meet on the waters of the Sky River (the Milky Way) on the bridge formed by the wing of a magpie.[8] It is the emperor of the sky who, long ago, had to make this river run in the middle of the sky in order to separate the stars in love, whose passion had driven them to laziness; the Herdsman was abandoning his flock and the Weaver her piece of work. However, moved by their grief, he allows them to meet once a year, during this summer night.

On this July 7, seven people,[9] three for the Weaver and four for the Herdsman, gathered on this "bridge of magpies" between the Orient and the West symbolized by Yamauchi Tokuryû, with the same ambition: to give the lie to Kipling's saying.[10] This chapter, in its own way, is its echo.

§ 31. The liberation of the logos

First of all, what is a lemma? The venerable *Lalande*[11] which by the way has no entry for *logos*, defines the term as follows:

> **LEMMA**, G. λῆμμα, what is taken (for granted), hypothesis; sometimes thesis. Is said particularly of the premises of a syllogism (*Topiques*, VIII, 1; 156a21). D. *Lehnsatz*;[12] E. *Lemma*; I. *Lemma*. A preliminary proposition whose preliminary demonstration is necessary to show the main thesis one intends to prove.

As for *logos*, there is no reason to restate all the definitions of this polysemic term, which since Heraclitus has stimulated European thought. The *European Vocabulary of Philosophies* lists possible translations:[13] "discourse, language, tongue, words, speech, rationality, reason, intelligence, foundation, principle, motif, proportion, calculation, connection, relation, tale, thesis, reasoning, argument, explanation, utterance, proposition, sentence, definition, account/tale ...". Following the article "Logos" that Anne Hourcade wrote for the *Dictionnaire des concepts philosophiques* (*Dictionary of Philosophical Concepts*) edited by Michel Blay,[14] let us remember the three main current definitions: word; thought or faculty of thought; reason in the sense of principle. The term is derived from a Greco-Latin root *LEG* (to pick, to choose), hence the Greek verb *legein* which first meant "to gather", "to count", and then "to say, to tell" and from there engendered the substantive *logos*. In Yamauchi's view, something has happened between Ionia and Greater Greece in the time of Heraclitus and Parmenides, which eventually would lead the logos to free itself from the constraints of being. I would rather say the constraints of existence, but the fact is that in Japanese those two notions are inseparable.[15] While Heraclitus (576–480 BCE) in Ephesus was the first to speak of the logos, Parmenides was the first to speak of being. The contrast here is between logos and being. Logic was at first the deployment of being, because for Parmenides (544–450 BCE), the father of the principle of identity (being is being) as his disciple Zeno was of contradiction,[16] Being and thought become identified. As the no less famous than obscure fragment VIII, 34, testifies:

> *Tauton d'esti noein te kai houneken esti noêma*
> To think and thought, are identical to being.[17]

This is usually compared to fragment III which clearly states that "indeed to think and to be is the same thing" (*to gar auto noein esti te kai einai*). Arnaud Villani has recently proposed an interpretation that is fascinating from a mesological point of view:

> In that case, which seems to be the solution to the enigma, fragment IV does not primarily deal with the *representation* of absent things by thought that renders them present, but above all with their pronouncement, even when mute, as thought words, that is, "understood" words. Continuing to work regressively, he ends up thinking that thought and being are the same *because both meet in words*. Without words, there is no being and no thought either. Even a thought that believes it is purely representative, *speaks, makes sentences*. This is Parmenides' beautiful discovery, well before the "sensible certainty" and the conception of language as universal in Hegel.[18]

This perspective fascinates me from a mesological point of view because, if being and thought "meet in words", that is simply the principle of trajection of the subject S (being) and the predicate P (thought on the *subject* of being) in the S/P reality. However, when it comes to knowing *who pronounces* this thought-being, a chasm opens up that the humanimal,[19] the *zôon logon echôn* (animal having speech), would be wrong to neglect. This book will end in facing this chasm; but in the meantime, let us get back to Yamauchi's ideas. He points out that two centuries later the dispute between Heraclitus and Parmenides would be resolved by Aristotle, leading to the emancipation of the logos and the creation of a formal logic based on the three principles of identity (A), contradiction (non-A) and the excluded middle (the impossibility of being both A and non-A).

Yamauchi schematizes how the "deployment of the logos" unfolds (pp. 40 ff.). In the first phase, being and the logos merge. The world of diversity, based on the privation principle (*sterêsis*), makes difference possible (to have or not to have such or such a character). In the second phase, that of opposition (*enantia*), affirmation and negation deviate from being, which however subsists as such. The same and the other are antagonistic, and relative to one another. For example, woman and man are of the opposite sex (rather than different, but their relativity presupposes the being of the one as well as of the other). This is no longer the case with the contradiction principle (*antiphasis*), the third phase in the deployment of the logos. Here, two contradictory terms are incompatible and only one of the two can exist. The logos now has free rein and transcends being. Even though the world of reality is both being and non-being, the world of the logos will categorically exclude their coexistence. The principle of contradiction will now structure logic. In Hegel's spectacular reversal, it becomes the dynamic of being itself. Contradiction destroys being, but by this very fact being moves, is deployed and rises up. It is the principle of being.

For Yamauchi, Hegel's reversal of the principle of contradiction is both the outcome and the summit of the deployment of the logos. However, Hegel did not violate the third principle of formal logic: the excluded middle. From even before Aristotle, and until today, it has never ceased to rule the logos.[20] And yet, oriental logic, outlined in *Treatise of the Middle Way* (*Mādhyamakashāstra*) attributed to Nāgārjuna (2nd–3rd century) which is the basis of the doctrine of the School of the Middle Way (*Mādhyamika*), or more widely the Buddhism of the Great Vehicle (*Mahāyāna*), overturns this third principle. To say that there is no logic except for the logos, shows the arrogance and abuse of power of western thought (p. 6). There is another logic in oriental thought, the lemma, that is not only an ideal-type of the cultures of the Orient, but "a wing of the global system of thought".[21] It should be possible to complete that system by acknowledging the lemma, which is what *Logos and Lemma* proposes.

We will shortly examine how Yamauchi develops this argument (§ 33); but first I would like to return to the idea that, with the principle of contradiction, the logos gave itself free rein with respect to being, or as I would rather say, with respect to existence. It seems to me that such an idea has some link to one of the characteristics of the Japanese language we have already seen in § 4: in that language you cannot say "Hanako is sad", you have to say "Hanako seems sad". In other words, you are forced to connect the utterance to the concrete existence of the speaker, in this case you yourself, who are not Hanako and therefore cannot say anything about her inner life.[22] By contrast you cannot help noticing that in English on the other hand, like in all major European languages, you can say "Julie is sad". You can do so because in those languages, the logos has free rein with respect to existence: even though you are not Julie and therefore strictly speaking incapable of saying that she is sad, the logos (your native language) allows you to get out of this concrete incapacity. In short, with this most banal utterance, you can transcend existence!

Furthermore, the same tendency towards abstraction is even more evident in Descartes' famous utterance: "I knew then that I was a substance ... which in order to exist, has no need of any place, nor depends on any material thing" (§ 1). In this case, the speaker is indeed situated in his utterance, but that couple is nowhere: he has taken himself out of any existential circumstance, exactly what the modern scientific view will do. The fact that such a linguistic arrangement generated formal logic, science or the *cogito*, is something that cannot be traced back to causality, but certainly to history and its contingency. To say the least, it is one aspect of the concrescence of the western heritage; to insist on the contrary would be pure abstraction, which is precisely what the logos authorizes. If it gives this authorization, isn't it because it excludes the excluded middle?

§ 32. The banishment of the third genre

First, speaking as a moderate follower of Sapir–Whorf (§ 6), I'll reaffirm that I place the link between language and thought under the sign of concrescence and contingency (from *contingere*, to co-touch, from *cum* and *tangere*), and not of causality. Sanskrit is an Indo-European language, but it is in Sanskrit that the *Treatise of the Middle Way* was written, which according to Yamauchi, presents thought derived from the lemma and not from logos. Is there a contradiction? No, because this link is not simply an abstract question of the logos but deals with a concrete, complex question of milieu and a history that goes way back into the past, even before the pre-Socratics. Ever since the 2nd millennium BCE, Vedic hymns have revealed a consciousness of language whose principles are alien to the logos. The short introduction to Sanskrit by Filliozat for instance includes the following strophe:

> The gods engendered the goddess Speech (*vācam*). Creatures of all kinds speak it. May this lovable Speech, the highly praised cow giving us its milk of strength and sap, come near us.[23]

The "speech" (*vāc*) in question is ritual. Its religious value puts order into the world, cosmizing it. It is derived from an Indo-European root, *WEK* which indicates the emission of the voice, which also engendered the Greek *epos* and the Latin *vox*, and gives us words like *epic*, *épopée*, *voice*, *word*, *advocate*, *revoke*, etc. Human realities to the core … However, what does the hymn in question tell us? That all creatures speak it! That all creatures are *zôa logon echonta* – animals having speech, as if such a thing could be thought in Greek, outside of fables …

The distant premonition of what biosemiotics will bring to light at the end of the 20th century, namely that the transmission of meaning is co-extensive with life (§ 27), not only puts the human being closer to other living beings (which undoubtedly has some connection with the idea of transmigration), more particularly it also refuses to abstract speech from all other phenomena;[24] in other words, preventing the logos from having free rein and betting on concrescence, the going-together-with of words and things.

Forging ahead recklessly, I will compare this strophe to one of the most famous stanzas of the *Treatise of the Middle Way* (II, 8), in Guy Bugault's[25] translation, which says:

> First, he who walks, does not walk, no more walks than he who does not.

This "he who walks does not walk" *gantā na gacchati*,[26] is precisely what the logos loathes. For good reason, at least given the sovereign dimension it has assumed. However, what about the phenomenon of walking considered as a whole? In other words, reality? Can you distinguish, separate, the agent of movement from the movement itself? No. The apparent absurdity of the paradox, "he who walks, does not walk", is only defined as such when language abstracts from the concrescence of the phenomenon of walking, a subject (the walker) on the one hand, and a verb (to walk) on the other hand; then, similarly, the logic of a subject on the one hand (the walker), and on the other hand a predicate (to walk); or in other words, the parent structures of the logos, SVC[27] and SP (§ 4), having free rein. This is the "logico-grammatical fiction" (Bugault's phrase, p. 55) which Nāgārjuna wants to unravel. Isn't it obvious that his intention is related to the aforesaid position: refusing to abstract speech from all the other phenomena, and betting on reality's concrescence?

To say that the logos separates (analytically deconstructs) reality's concrescence, is also saying it unravels the going-with of words and things, in other words it de-symbolizes.[28] In the course of history, this tendency ended up generating the MOT and modern de-worlding; this is especially the case in the disjunction by capitalism of the worker and work, which is strictly homologous to the one Nāgārjuna denounced, between the walker and the walk. To quote Alain Supiot:

> Work is inseparable from the one who works, and this is why until capitalism, its legal status always depended on the personal status of the worker. For work, and not its product, to be treated like the object of a contract, in other words, for a labor market to be established, the couple *psyche/soma* must be disjointed in order to introduce property relations between man and his own body.[29]

However, at what moment did this disjunction occur? When Parmenides proposed the principle of the identity of being? This is the current view, which Villani has radically contested recently.

> What if Parmenides is not the traditional figure we imagine him to be? What if he were not the logician of the principle of identity we think he is? What if Being were a "holding together" and the Sphere a well-kept world, respecting the Truth of the Goddess without excluding the

> Sensible in any way? … I maintain that Parmenides' intention, even if it only appears between the lines and in suggestive winks, is to familiarize us with a dual system (which is certainly losing steam, as its grammatical case is disappearing); it includes elements that only come in pairs and whose effect is *to reinforce unity*. Both ethnography and philosophy call this *symbolic thought*. I even think that Parmenides as well as Heraclitus deal with a similar *struggle* that is already lost. They want to give a warning, one last time, at the moment when the *Logos* is definitely overtaking the *mythos*, that the symbolic is not outdated and should remain a major component of thought. However, the unilateralism brought in its baggage by the *Logos*, triumphant in its radiant youth, is *the mother of all perils* if its hegemonic tendency cannot be contained in the imperative limits of the symbol.[30]

Villani's thesis says that "the Pre-Socratics are not the stammering embryo of Platonic and Aristotelian philosophy. They constitute another beginning of western philosophy" (p. 5), and he goes so far as to compare them with Taoism.

Whatever the case may be, the logos prevailed, leading to the impossibility of thinking the symbolic, where the dual is one, where A is always also non-A, and where words have meaning because they go concretely with things. In the *Timaeus*, this powerlessness shows clearly with respect to the *chôra*, which concerns us here: in European thought this notion is the ancestor of the problematic of the milieu, and hence of mesology.[31]

In the case of the *chôra*, the least one can say is that the Platonic discourse is not clear. The reasons are probably twofold and actually contradictory. The text of the *Timaeus* does not overcome this contradiction and determines the *chôra*'s fate in European thought for centuries. In short, the latter will forget the *chôra*, and forget the question "why do beings need a *where?*" in order to reach a clear definition of the notion of *topos* that Aristotle substituted, i.e., to answer the question: "*where* are beings?" As we will see, this precisely means excluding the *chôra* from the question of being.

While in Plato's reflections on the *chôra* in the *Timaeus* this foreclosure has not yet taken place, by contrast his ontology, whose principle is the self-identity of "real being" (*ontôs on*, i.e. the *eidos* or *idea*), excludes any logical grasp of the notion of *chôra*, which mysteriously eludes the principle of identity. It is so elusive that Plato gives no definition at all and confines himself to metaphors that are moreover contradictory. Here he compares it to a mother (*mêtêr*, 50d2), or to a wet-nurse (*tithênê*, 52d4), one could say a womb, but elsewhere to the opposite of a womb, that is, an imprint (*ekmageion*, 50c1). The *chôra* is both imprint *and* womb with respect to what Plato calls *genesis*, i.e., the becoming of beings in the sensible world (*kosmos aisthêtos*). In the onto-cosmology of the *Timaeus* genesis does not partake of real being but is only its reflection or image (*eikôn*).

Both imprint *and* womb, a thing and its contrary at the same time, the *chôra* literally has no identity. One cannot even imagine what it is. Plato recognizes that such a thing is "hard to believe" (*mogis piston*, 52b2), and "we see it as in a dream" (*oneiropoloumen blepontes*, 52b3) but he insists it exists: in the ordering (the cosmization) of being, there is simultaneously and immediately, real being, its projection into existing beings, and the milieu where this projection is achieved concretely as becoming, in other words the *chôra*. In the Platonic text: *on te kai chôran te kai genesin einai, tria trichê, kai prin ouranon genesthai* (52d2), "there is being and milieu and the existent, all three triply, born before the sky". (That is, before the ordering of the *kosmos*, which the *Timaeus* identifies with *ouranos*.)

The phrase "all three triply" is problematic. *Trichê*, means "in threes, in three parts, in three different ways". Why this doubling with *tria*, the neuter of *treis*, which just means "three"? Of course, forget about seeing here any foreshadowing of the Holy Trinity. Even if one could possibly see in the historic incarnation of the Father in the Son some analogy with the projection of the *idea* in the *genesis*, the *chôra* itself has nothing to do with the Paraclete. It is much too physical! Plato's comparison with a "wet-nurse", *tithênê*, is a word that has the same root as the French *téter* (to suckle) or the English *tit*. The *chôra* is apparently breastfeeding the new-born, that is genesis, whose primary meaning is "birth". Moreover, in 50d2, the *Timaeus* compares the absolute being to a father, the milieu to a mother and becoming to their child. That child has to be fed!

We have here two leads for our research. The first will be to clarify the idea of maternity that *chôra* connotes. The second, to further investigate the theme of the trinity of being, milieu and becoming.

As for the first theme, going beyond the images themselves in Plato's text (the mother, the wet-nurse), we must return to a more general and more concrete sense the *chôra* may have had in the Greek city. For the citizens of the *polis*, beyond the ramparts of the *astu* (town), the *chôra* was the nurturing countryside whose hills they saw every day, covered with wheat, vines and olive trees. It was the place that daily provided the earthly nourishment allowing them to survive. In that world, no *astu* without *chôra*!

The word *astu*, the urban center of the *polis*, has the Indo-European root *WES* which means "a stay, to remain". This root can be found in the Sanskrit *vasati* (he stays) or *vastu* (site). In German it became *Wesen* (to be, nature, essence), *war* and *gewesen* (forms of *sein*, to be); in English *was* and *were*, (forms of *to be*.) In the case of a Hellene like Plato, the *astu* is essentially the site of being (Castilian would say: *la estancia del ser*): however it cannot exist concretely without the nurturing milieu of the *chôra* surrounding the *astu*.

This concrete relation, or growing-together, has best been exemplified by Russian archeology in the Crimea; this land's history (in this case Tauric Chersonese) allowed for the structures of the *chôra* characteristic of a Greek city in antiquity to clearly subsist in the landscape.[32] The white marble stone slab of the "Oath of Chersonese", which more or less dates from Plato's time, reveals that each citizen made a lifetime commitment to defend this territory: "I swear by Zeus, Gaia, Helios, the Virgin (*Parthenos*), the gods and goddesses of the Olympus and the heroes who own the polis, the *chôra* and the fortified posts ...".[33]

The theme of the trinity in which the *chôra* occurs becomes more complex since it is neither the absolute being (the *ontôs on*), nor the relative being (the *genesis*). Neither this nor that, but something else, which, Plato tells us, is a "third and different genre" (*triton allo genos*, 48e3). Other than what? Other than the two *eidê Timaeus* has just been talking about, i.e., the *on* and the *genesis*. *Eidê* is the duel of *eidos*, species (word derived from the Latin *species*, view, sight, aspect), a form in the mind that we see as an idea. The word is related to the Latin *videre*, to see and to the Sanskrit *veda*, I know, *vedah*, aspect. All these words come from the Indo-European root *WEID* indicating the view associated with knowledge. For Plato, the *eidos* or *idea* is the real being, the absolute; this is why we speak of his "idealism". However, clearly the *genesis* could sometimes pass for *eidos*, or being. This does not change the fact that for Plato there are two ontological orders: the first is a "type of being of the Model, which is understandable and always identical to itself" (*paradeigmatos eidos ... noêton kai aei kata tauta on*, 48e4), the other one, "the second type, the copy of the Model that is generated and visible" (*mimêma de paradeigmatos deuteron, genesin echon kai horaton*, 49a1).

The third type is the *chôra* which is certainly also a "presence" (*paron*, 50c7) like the other two, but is "difficult and dimly visible" (*chalepon kai amudron eidos*, 49a3). Can we actually say it is an *eidos*? That remains to be seen. It can only be approached through metaphors, some of which we have seen, but which ultimately do not yield any definition. All we have is the designation of the *chôra* as "an invisible [an oxymoron, since the *eidos* is an aspect, presupposing vision] and formless species [another oxymoron since the *eidos* is a form] which receives everything (*pandeches*) and partakes of the intelligible very aporetically" (*metalambanon de aporôtata pê tou noêtou*, 51b1).

The "third genre" of the *chôra* thus remains an aporia: an insuperable obstacle for the logos which, confronted with this third genre, is left without resources and cannot go on (*aporei*).[34] Why not? Obviously the question is linked to the capacity of the logos itself, and therefore with logic. Its elaboration will have to wait for Aristotle, but the three principal actants are virtually there: the principle of identity (A: the Model which is always identical to itself), the principle of contradiction (non-A: the copy which is not the Model), and the principle of the excluded middle; the *chôra*, neither the Model nor its copy, but the "third genre", neither A nor non-A which remains incomprehensible, excluded by the logos.

The *Timaeus* went no further. We can conclude that with the *chôra*, i.e., the subject of this essay, Plato foreclosed a realm of thought that could have been developed but which remained unexplored because of him. To quote Alain Boutot: "By transforming in and through his idealism the primitive *chôra* into an amorphous milieu receiving all bodies, Plato decisively hides the original dimension of spatiality as region (*Gegend*) that was perceptible in earlier thought."[35]

Why couldn't the idealism of the *Timaeus* grasp the regionality (*Gegendheit*[36]) of the *chôra*? Why did the "third genre" remain aporetic? In the framework of the ordering of thought (which is cosmisized) with Plato, and instituted in the *Republic*, we are done with myth, in other words with the symbolic, at least *logically*, that is, according to the logos. From now on, one must "be or not be", but not both at the same time as in symbols where A is also always non-A. This is why poets, the people of symbols, were banned from the Platonic Republic: *ho gar logos hêmas hêrei*: "the logos imposed this duty on us" (*Republic*, I, VIII, 609b3).

§ 33. The third of the two truths

Could the lemma make it possible to "grasp" (*lambanein*, hence *lêmma*) and concretely *comprehend* what the logos foreclosed by its abstraction? *Logos and Lemma* intends to show this in a dense discourse of close to 400 pages which we obviously cannot summarize here. We will consider only what directly concerns mesology;[37] indeed, Yamauchi's argument is to a great extent a commentary on Nāgārjuna's *Treatise of the Middle Way*. To be sure, the use of the word "middle" (*mādhya*),[38] does not correspond *ipso facto* to the subject of mesology. The *Treatise of the Middle Way* is not a mesology, but there is a connection that sheds light on mesology from the very first stanza:

> Without anything ceasing or occurring, without any annihilation or eternal life, without unity or diversity, arrival or departure, such is the conditioned coproduction.[39]
>
> *A-nirodham an-utpādam /an-ucchedam a-çāçvatam /an-ekārtham a-nānār-tham / an-āgaman a-nirgamam / yah pratītya samutpādam.*[40]

We see that the text begins with a succession of four pairs of denials, what the Chinese gloss calls the "eight negations" (*ba bu*, Japanese *happu*). Bugault (p. 35) interprets them as a rejection of dualistic thought and mental dichotomies. Yamauchi (pp. 71 ff.) underlines the fact that the twofold negation (neither A nor non-A, neither affirmation nor negation) is placed right at the beginning. He believes it shows the particular importance of double negation in the *Treatise*. This makes it the key to Nāgārjuna's reasoning, based on the tetralemma.

– What is the tetralemma?

– A reasoning in four lemmas, usually presented in the following sequence: 1. affirmation (A); 2. negation (non-A); 3: twofold affirmation (both A and non-A); 4. twofold negation (neither A nor non-A). According to Yamauchi, such a sequence leads nowhere. It is an indefinitely repeated loop, without any development of thought. For him, the double negation is decisive and should occupy the nodal position of the third lemma, which then makes possible the fourth, now the double affirmation. It is literally the opening, the starting point of the Great Vehicle Buddhism, founded by Nāgārjuna.

As the double negation negates both affirmation (A) and negation (non-A), it is absolute negation. In the tetralemma, the third lemma introduces a break, a decisive gap with respect to the first two lemmas. While the latter are used in a worldly or profane context, (*vyavahāra*, Japanese *sezoku* 世俗), the third and fourth pertain to what the Great Vehicle calls the supreme truth (*paramārtha*, Japanese *shôgi* 勝義). Starting right away with the eight negations, the *Treatise of the Middle Way* focuses on this supreme truth, beyond the profane world. Worldly logic only concentrates on the double meaning of the first two lemmas, either A or non-A, without a third term that would be both A and non-A. It is the principle of the excluded middle which the *Treatise* does not ignore in the least, as is clearly evident in stanza II, 8, mentioned earlier, (§ 32). Here is the whole stanza:

> First of all, he who walks does not walk, nor does he who does not walk. Which third party, other that the agent of the walk and his non-agent could then walk?[41]

The issue here is to *transcend* and not ignore the principle of the excluded middle. To get to the supreme truth, one must first go beyond that principle; this is precisely, according to Yamauchi, what the double negation achieves, as the absolute negation making possible the double affirmation, i.e., the fourth and last lemma.

– "Profane world", "supreme truth", this is religion, not logic!

– Yes indeed, the *Treatise* of the *Middle Way* deals with religion. The dedicatory stanza ends with the following words: "He who taught us this [i.e., conditioned coproduction], the perfectly Awake One, the best of instructors, I salute him" (*ibid.*). This is obviously addressed to Buddha, to whom the whole treatise is dedicated. It is not a work of logic, but of religious thought, which argues for or against religious positions even as it rests on a rigorous mental framework: the tetralemma.

– In what way does this religion, Buddhism, concern mesology?

– Through the tetralemma, since it allows for the overcoming of the principle of the excluded middle.

– But this is no longer logic.

– It is indeed no longer *logic* because the logos stops before the wall of the third lemma. It cannot pass and we have an impasse. However, it cannot pass in the domain it seized when giving itself free rein, abstracting from concrete reality. In the domain of the logos, there is no way to overcome the double aporia of the third and fourth lemmas. Both are absurd. Let us return to stanza II, 8, the one about the walker who does not walk. As we have seen, it means that the logos is incapable of grasping the unitary reality of that phenomenon because it necessarily breaks it down into subject and predicate. The idea goes well beyond this particular example. From a mesological point of view, it implies that the logos cannot grasp the trajectivity (S/P) characteristic of every phenomenon, because the latter implies both the thing and our existence, the instance predicating the being-in-itself of a certain thing. Instead, the logos abstracts the real from our existence and dichotomizes it into an agent on the one hand and an action on the other, or into subject and predicate. In this case, we are not dealing with logic but with *lemma thought*, based on the tetralemma. This is clearly illustrated by the *holds* of the ecumene (§ 26) *that are literally lemmas* (from *lambanein*: to take); to be more precise, fourth lemmas, because they are active and passive at the same time, imprint and womb. They are *syllemmas*[42] (from *sullambanein*, to take together), that grasp together and thus literally *com-prise* A and non-A.

– Are you sure this is what Yamauchi means?

– No, this is not what he means; this is what it means for mesology. From Yamauchi's point of view, which corresponds to the orthodoxy of the Great Vehicle, there are two truths, one profane (lemmas 1 and 2) and the other supreme (lemmas 3 and 4). For mesology, there is only one, *the third truth*. From that point of view, the tetralemma makes it possible, not to reach a "supreme truth" transcending the profane world, but on the contrary to fully understand profane reality, that is, simply the reality *in which we live*: our milieu. It is eco-techno-symbolic in as far as things are trajective; this means nothing can be reduced to either the simple identity of an S object, that would be A or non-A, or that of a predicate P, which in turn would be A or non-A. A concrete thing is always S/P. It is *S as P*, that is, it cannot be reduced to *either* an S *or* a P. It is neither S nor P, neither A nor non-A, in other words neither a simple object nor a simple representation of that object. "Neither of the two terms", that is a double negation. It is the third lemma, the one according to Yamauchi that makes possible the fourth (the double affirmation). It is precisely because the thing cannot be reduced to a pure object (S) or a pure predicate (P) that it becomes concrete reality: S as P, that is, *both* S and P. Not only is it both substance (S) and non-substance (P), both A and non-A but in opposition to dualism, it also presupposes a third term, which is the predicating instance for which and through which S and P are possible: in other words, an existence (our own in the first instance) embedded in its milieu. It is the fourth lemma: the double affirmation. It is the reality which, like the fourth lemma, *com-prises* (takes together) our existence and that of things, precisely what dualism, the son of logos, decomposes into subject and object, subject and predicate, thing and word, signifier and signified, etc.

– It vaguely reminds me of the difference between the binary semiology of Saussurians and the semiotic triads of Peircians …

– Yes, and it is no coincidence that biosemiotics was inspired by Peirce and at the same time by Uexküll's *Umweltlehre*, in other words by mesology. We will have more details in the following chapter, but we can already express this in less scholarly terms. In a human milieu, everything that surrounds us has necessarily, and simultaneously, a triple dimension that is ecological, technical and symbolical. The *ecological* dimension means that all things depend on one another; none of them can be abstracted from the ecosystems of the biosphere. Even the most artificial must find here their raw material, and cannot fail to leave their mark. The *technical dimension* means that all things, one way or another have a meaning that depends on our technical systems; this meaning thus evolves in the course of history. Yesterday we used the wind to grind our grain; today it could turn us away from nuclear power. The *symbolic* dimension means that all things, in spite of their physical identity, may be different depending on our symbolic systems, starting with our languages: the same displacement of air that here becomes "wind," is *el viento* there, or *kaze*, etc. All this means is that things are always at the same time, indissolubly, S and P, substance and non-substance, A and non-A, word and object represented by the word. They necessarily embody the fourth lemma of the tetralemma, which remains an aporia for the logos, which in turn strikes back against language by separating it into signifier and signified, etc., in other words the symbolic and *a fortiori* the medial. The logos can only deal with abstract objects (S, *in principle*), not concrete things (S/P, *in fact*).[43]

– Briefly, the concrete reality of human milieus is based on the tetralemma and belongs to lemmic thought, not logic. Concreteness is lemmic, abstraction is logical. Is that the "third truth?"

– Yes, it is the "third" in as far as it still recognizes the existence of the first two, the "profane" truth, based on lemmas 1 and 2. Along with the principle of the excluded middle, it makes possible the abstraction characteristic of the rational constructions that found logic, science and modern technology. That third truth also allows for what in the Buddhism of the Great Vehicle is called the "supreme truth" based on lemmas 3 and 4, because this *dimension really and universally exists as soon as one admits the phenomenality of phenomena*, i.e., that things are at the same time, and *in reality*, both what they are (S) and the way in which they appear *to us* (P). This necessarily presupposes the symbolic, because there is a symbol as soon as an existent, no matter what it is, enters into a relation with any object (S), and interprets it as something (S/P). The existent, and thus the existence of *our* sensible world, not of some universe alien to us, is the third symbolizing aspect which one excludes and forecloses when separating S and P, and stopping at this duality as the logos does with the principle of the excluded middle. The symbol is equally necessary because there is mediance, meaning that S/P says the same thing in every medial body, beyond the individuality of each member. There is universal mediance as soon as there is existence, because there are necessarily species.[44] What is individual, whether object or subject, and the individual human subject in particular, remains just an abstraction when separated from concrete reality. There is no human except in a milieu and as society, no society except in the ecumene and as humanity, but there is no humanity either, except as species, and no species finally, except as long as there is life on Earth in the biosphere, etc. Part III of the book will deal with this issue.

– OK. In the meantime, that is a lot of "as such and such", and you have said nothing about this "conditioned coproduction" *pratītya samutpāda*, invoked by Nāgārjuna in the very first sentence of the *Treatise of the Middle Way*, and which he clearly associates with the third lemma.

– Precisely! From the mesological point of view, it is directly linked to this "*as such and such*", *which is not a* "*because*". The question here is not of causality (A brings about B), but of concrescence (A and B going together) and even of *co-suscitation* (A and B bring each other into existence).

– I remain skeptical …

§ 34. Short of the mystical leap

The Sanskrit phrase *pratītya samutpāda* today is usually translated, as Bugault does, by "conditioned coproduction". Roger-Pol Droit introduces it in the these terms:

> It is a complicated expression. It is usually translated by "conditioned coproduction" or sometimes "dependent production," "interdependence".
>
> The Sanskrit word is composed as follows:
>
> – *pratî* as, according to;
> – *ya*, the gerund of "i", to go,[45] going as, according to;
> – *sam* the mark of convergence;
> – *utpâda* means "production".
>
> We are dealing here with a technical term which can be literally translated as "convergent production going according to". It refers to a central theoretical development of the Buddhist doctrine, whose main principles we will recall here.
>
> The basic formulation is in the Pâli Canon of the Theravâdin, dating from the 1st centuries before our era. It is probably the first formulation of a law of the conditionality of phenomena, but also, as Guy Bugault pointed out, of the very idea of functioning according to.[46]

From a strictly grammatical point of view, the commentators' analyses are quite similar, but they can lead to very different interpretations. Yamauchi (pp. 145 ff.), citing the *Prasannapadā* by Candrakīrti (7th century) and

Yamaguchi Susumu's commentaries on it, mentions that the word *pratītya-samutpāda* is composed of *pratītya* and *sam-utpāda*, *pratītya* composed of *prati* and *itya*, where *iti* is *eti* (to go)[47] and *prati* to reach; *prati* + *iti* forming the meaning "to go and reach"; *sam* (an ensemble) composed with *utpāda* (to occur, to appear); hence *pratītya-sam-utpāda*, i.e. this means that

> things happen, through co-awaiting by cause and relation; in other words, they co-suscitate.
> *shohô ga in to en to ni sôdai shite shôki suru no ga engi de aru*
> 諸法が因と縁とに相待して生起するのが縁起である
>
> (p. 146)

– Co-awaiting? Co-suscitation?

– I'll get back in a moment to the concept of co-awaiting (*sôdai* 相待). Co-suscitation is my way of reading the sinographs 縁起 (Chinese *yuanqi*, Japanese *engi*), which in eastern Asia have become the canonic translation of the Sanskrit *pratītya samutpāda*.[48] The first (縁) means "link, linked", and the second (縁) "to arouse, to give rise to". Hence my translation of "co-suscitation", in other words, the mutual effect that things have on each other, including through feedback. This goes much further than causality, without excluding the latter in any way; this is where we encounter the other concept, *co-awaiting*.

Yamauchi introduces that concept[49] in a commentary on the absolute negation represented by the third lemma. On page 102 he says:

> The position of the Middle Way [*Madhyamaka*, Japanese *Chûgan* 中観, i.e. professed by the school founded by Nāgārjuna] does not consist in stubbornly negating everything a propos of everything. It wants to negate each of the two terms of what is connected.[50] Concerning birth and death, it negates birth by negating death but does not negate either birth or death.[51] Birth and death are changes, changes that are realities and cannot be negated. To say that movement, uniting being and non-being, cannot exist,[52] does not prevent the wind from blowing nor water from flowing. You can argue about these realities as much as you wish, you cannot negate them. Negating them would really be nihilism. The negation of the Middle Way negates birth by negating death, but this does not mean it negates birth and death. If it negates birth as it negates death, it is rather to affirm non-birth as non-death. Birth and death correspond with each other, and what negates this correspondence (*sôtai* 相対), is the absolute negation. The absolute (*zettai* 絶対) is precisely what stops correspondence (*zessuru* 絶する). However, birth and death are moreover co-anticipating (*sôdaiteki* 相待的). Birth is birth in terms of death, death awaits birth to become death. If it is abtention (*zetsudai* 絶待)[53] to negate this co-awaiting, then the non-birth-non-death is indeed abtention. Abtention is the void (*kû* 空). If

> the absolute is the negation of correspondence, the negation of co-awaiting must be the void.[54] In the next chapter I will discuss in further detail the difference between correspondence and co-awaiting. But we must know right from the start that the absolute belongs to the logic of the logos, while abtention pertains to the point of view of the lemma.

We see that Yamauchi introduces the concept of the void, one of the central themes of Buddhism. His argument introduces it because the domain opened up by the third lemma is also for him, as for Nāgārjuna, the domain of the supreme truth. Mesology does not consider this domain to be related to the supreme truth, but simply a part of the profane reality surrounding us (§ 33). The question is what could correspond to this "co-awaiting"?

Let us first point out that Yamauchi's argument is not limited to recondite concepts such as those I just translated. There are plenty of examples that are more easily understood, for instance, to explain co-awaiting, we can look at parents and children. The parent cannot be parent without the child, the child cannot be a child without a parent. Both exist as such only in this relation; in other words, to be what they are, they must mutually anticipate each other. According to Buddhism, neither of the two terms in this relation has it own nature or being-in-itself (*svabhāva, ātman*, Japanese *jishô* 自性). The one acquires its being-such only at the moment when the other also acquires its being-such. For the same reason, everything is always "in awaiting" of the other to be what it is; it becomes such and such by the "tension towards".[55] This is co-awaiting. For Yamauchi, as for the Great Vehicle, it pertains to the supreme truth, whose central motif is the void. According to Yamauchi, it appears first in the double negation operated by the third lemma; in the example above, it means to negate the parent (*neither* A) and to negate the child (*nor* the non-A), since neither one nor the other exists in themselves. To believe in such existences is only the illusion of profane truths.

There is however a problem with making relativity absolute; for mesology it means making the phenomenal absolute. If everything waits for everything to become something, doesn't it follow that nothing ever happens? However, things do exist and events are happening! Buddhism solves this problem with the help of a concept that Yamauchi uses in turn: propping (*niśraya*, 依止 – Japanese: *eji*, also read as *eshi*). Girard[56] gives the following translations: support; to lean on a master, a virtuous person; to live with a master. He adds this quote from the *Mahāyānasūtrālamkāra*:[57] "It is because they do not have their own nature that [all the *dharma*] stack up/ The anterior is the support of the posterior."[58] In short, one relation settles on the one that precedes it and all settle on each other mutually, without needing to substantify the terms.

This however only postpones into infinity the answer to the question of the origin of things. This is why Buddhism, like all religions must have absolutes (the void, supreme truths …) to be able to say something to people, in both senses of the word "to say": to tell them something and to attract them somehow. All these religions must *absolutely* make the mystical leap toward the absolute, whether substance or non-substance, being, void, nothingness, whatever, as long as it transcends the profane. Mesology cannot follow Yamauchi, or Buddhism or any other religion, in this mystical leap beyond which certain truths become supreme. Mesology will even refute that perspective: it claims that religions, mystical leaps and "supreme truths" can be explained by the inherent symbolic dimension in human milieus that always transcends the physical (at the level of the sensible world, that is).[59] Besides, mesology will never deny that we are dealing with realities, because they are in fact the S/P relation, but they are indeed *human* realities (§ 25)!

– Does this mean that mesology only pertains to the logos, and that the domain of the lemma will forever elude it?

– With concepts such as mediance and trajection, mesology claims to go beyond the alternative of the in-itself and for-oneself, the Earth turning and not turning. The purpose of *Logos and Lemma* shows a profound analogy with Husserl's intent in *The Earth does Not Move*: to place the field of the logos (where the Earth moves) within a deeper and larger field, called the archeo-originary (*Ur-Arche*) by Husserl and lemma by Yamauchi. For mesology, it is the trajective, or *lemmic* field of the co-existence and growing together of the in-itself, objectified by the logos, and the for-oneself necessarily established by the objective event that we exist. It is the field of total reality that finally can com-prise (*comprehendere, sullambanein*) without merging them, both object and subject, our existence and what science reveals us. This is the *syllemma*, the "taken-together" symbolized by the fourth lemma.

This claim is not the swansong of subjectivity faced with the tsunami of scientific objectivity, but on the contrary, the desire to fully take into account all the evidence provided by physics itself after three centuries of modernity. As Heisenberg expressed it best (§ 29), the method can no longer be separated from the object since the method transforms its object. This is the very principle of trajectivity! Or also of co-suscitation ...

– Yes, but between the Planck length[60] and what you are saying with your tale of Rousseau's dawn, isn't there a slight problem of scale?

– Precisely, this problem of scale removes the mystical leap I mentioned. The leap allows us to apply to every phenomenon, without distinction of scale, the principle of Schrödinger's cat, in other words the fourth lemma (the double affirmation or syllemma: both A and non-A), to claim anything and its contrary at the same time. For example, Nishida with his concept of "absolutely contradictory self-identity",[61] or claims that to make war is to negate war. This is what happens when one absolutizes what should not be absolutized, without attention to scale. However, since mesology deals with the ecumene, it is necessarily geographical, and takes from geography the primordial concern with scale. It concerns the *concrete* scale of phenomena that always go together, never in the absolute; not just in space but also always in time, which means *concretely*, in history.

While mesology's rejection of any mystical leap makes it agnostic, it however owes *Logos and Lemma* multiple insights concerning profane reality. To be sure it founds mediance and trajection neither in the absolute nor in the void, were it abtentive (*zetsudaiteki* 絶待的), but adamantly only in the concrete occurrence of phenomena in the ecumene. As I did not participate in the seminar of Shin Hirayu on Yamauchi's work (§ 30), and since no public traces[62] remain thereof, I will never know what he meant by the expression *mono kara koto e* 物から事へ, "from the things to facts" that announced his talk. I feel therefore free to imagine it could concern the passage of the *in-itself of things* (the Earth turning) to the *fact* that we exist, and therefore means these things happen *to us*, just as the sun rises *for us* (the Earth does not move). From Galileo to Husserl, as from Husserl to Galileo, this lemmic passage (*poros*), turned into a non-passage (an aporia) by the logos and hence an alternative (either the Earth turns or it does not turn), means trajection, the to-and-fro fully instituting reality. In that perspective alone, a number of concepts and arguments outlined in *Logos and Lemma* become paradigmatic. They illuminate the very problematic of mesology.

For instance, the concept of co-awaiting. If we exist and things exist for us, then the milieu and the symbolic (§ 33) exist too, just as there is simultaneously mediance of our animal body and our medial body. Not only are the things of our milieu effectively related to our existence and reciprocally, but they are also in co-awaiting, that is in tension with their others. This is because they are in the trajective field of reality (S/P) which also includes us. For example, Heidegger would not have disavowed the notion that the nail and the hammer are co-anticipating, because in that field (a milieu in particular, the ecumene in general) they co-suscitate each other with and through our existence. This is not limited to functionality (simply a question of the logos) because concretely, it presupposes and engages our existence, which *motivates* both the nail and the hammer. This is indeed trajectivity; however to speak of co-awaiting and co-suscitation in that connection, like speaking of lemma in connection with the *holds* on our milieu (§ 26), means to corroborate mesology with perspectives that are of far greater historical significance and have more authority than the few pages of a simple essay like this one.

It is the same with what Yamauchi calls "the logic of *soku*"[63] (I prefer to call it "the lemmic of the *soku*") to which *Logos and Lemma* devotes Chapter 11. *Soku* 即 is also read as *sunawachi*, usually translated as "that is". In Buddhism, according to Girard's definition,[64] *soku* expresses:

> the identity of two inseparable things that are like the front and back of a single reality. This is temporal immediacy, whether it be the absence of any distance between two facts, as for instance when

darkness disappears when light comes on, or whether two things are inseparable even when removed in time, – like the culture of the seven members of the perfect awakening after benevolence, or the painful retribution after a bad action.

For Yamauchi, the "logic of the *soku*" is what occurs in the passage from the third to the fourth lemma, in an immediacy both temporal[65] and spatial[66] pertaining to intuition, not dialectics. In short, *soku* expresses a relation between the two terms A and B, in which A is/is not B. It is the fourth lemma, the syllemma where A and non-A "are taken at the same time". Mesology sees here a kinship with trajectivity, where "S as P" means that S is/is not P, and vice-versa. In other words, S *soku* P; this pertains to the lemmic of the holds of the ecumene rather than to logic. Given the richness of the traditional thought in east Asia related to *soku*, it constitutes a fascinating source for further research into the problematic of milieus.

Take for instance the notion of propping (*eji*) discussed above. Detached from the void as well as from supreme truths, and brought back to the concrete scale of history and human milieus, this notion becomes what I call the *trajective propping*. This is what happens in trajection, when a new assumption is settled in the hypostasis of an earlier trajection: it is P′ propped up on S/P, which makes it into S′ and so produces (S/P)/P′, that is S′/P′, and so on (§§ 8, 15, 21, etc.). Literally, as for the *Mahāyānasūtrālamkāra*, "the anterior becomes the support of posterior," *qian wei hou yizhi* 前為後依止！

Clearly, this presupposes progress in time; concretely this means a history. However, isn't history, including the history of science, precisely the trajective propping of what is posterior by the anterior, in other words, of assumption by hypostasis? There is no need here to absolutize either S or P: they are co-awaiting, just *as*, and *because*, our animal body and our medial body co-suscitate each other to be what they are. We exist in that structural moment, which is our mediance, and that is sufficient. For sure, to stop there is to give up on the supreme truth and on an answer to the question of the origin of things. Nevertheless, the profane and agnostic use of the tetralemma does not exclude wider perspectives; it means to conceive of reality as the indefinite progress of a trajective chain, to effect a radical turn towards history, and together with the milieu, to make history into one of the two pillars of mesology. The third and last part of this book will try to show this.

Notes

1 YAMAUCHI Tokuryû, *Rogosu to renma*, Tokyo, Iwanami, 1974. Many authors read 山内得立 "Yamanouchi Tokuryû", which is also possible and more current. However, I prefer the reading "Yamauchi" which is specified at the end of the book in question.

2 *Mono kara koto e to yue to no mondai* 物から事へと故との問題. This was the title and probably the real theme of the seminar.
3 Those five thick volumes were published between 1944 and 1960. Yamauchi was certainly a seasoned Hellenist! A universal philosopher, he had started as a Germanist (he was Husserl's student) and he also read in the original text Chinese and Sanskrit classics on which *Logos and Lemma* is based.
4 A summary in French can be found on the Web on the site "Mésologiques: Logos et lemme" (Logos and lemma). Note of translator (July 2018): a French translation of *Logos and Lemma* by A. Berque is in press at CNRS Éditions, Paris.
5 Rudyard KIPLING, "The Ballad of East and West", included in Edmund Clarence STEDMAN, *A Victorian Anthology*, New York, Bartleby.com, 2001 (1895). While the rest of the poem shows that respect is possible between an English officer and an Afghan horse thief, nothing invalidates the idea contained in the first two verses that the Orient and the West cannot understand each other.
6 *Mono kara koto e to fûdo no mondai* 物から事へと風土の問題.
7 Ed. Laurence CAILLET, Paris, Publications orientalistes de France, 1981, p. 272.
8 Even though the morphology doesn't show it, the traditional Japanese expression *kasasagi no hashi* (read also as *kasasagibashi* or *jakkyô*), directly copied from the Chinese *que qiao* 鵲橋, is in fact heard as a plural: "the bridge of *magpies*". While retaining its original meaning, it means "happy union", and popular iconography associates it with marriage. In China, the magpie has symbolized married bliss for a very long time. (See Maurice-Louis TOURNIER, *L'imaginaire et la symbolique dans la Chine ancienne* (The imaginary and symbolic in ancient China), Paris, L'Harmattan, 1991, pp. 173–175.)
9 Of course seven goes with Tanabata, but it was also about playing, under the cryptomeria (Japanese cedar tree) of the Hida mountains, at "pure chats" (*qingtan* 清談) of the Seven Sages of the Bamboo Grove (*Zhulin Qi Xian* 竹林の七賢), who lived in China in the 3rd century. Those chats were one of the origins of what Yamauchi calls lemma thought. See my book *Histoire de l'habitat idéal* (History of the ideal habitat), *op. cit.*, § 7, on the period when the notion of landscape emerged.
10 Entitled *De chose en fait. La question du milieu* (From thing to fact. The question of the milieu), a selection of the presentations at this meeting, translated into French, was published in a special edition of the magazine of the Maison franco-japonaise, *Ebisu*, 49, Spring–Summer 2013 including Présentation (Presentation) by Augustin BERQUE; Les souvenirs de Monsieur Sôkejima (Memories of Monsieur Sôkejima) by EGUCHI Kumi; L'horizon de la logique lemmique (The horizon of lemmic logic) by KIOKA Nobuo; Tétralemme et milieu humain (Tetralemma and human milieu) by Augustin BERQUE; La raison des choses (The cause of things) by Frédéric GIRARD; La raison-coeur des co-suscitations paysagères (The sensitive reason of landscape co-origination) by NAKAMURA Yoshio.
11 André LALANDE, *Vocabulaire technique et critique de la philosophie* (Technical and critical vocabulary of philosophy), 7th ed., Paris, PUF, 1956 (1902–1923), p. 556.
12 Literally, this means "proposition" on which one "leans" (*lehnen*).
13 *Op. cit.* in § 2, p. 727.
14 Paris, Larousse/CNRS, 2006, pp. 479–480.
15 Here I translate *sonzai* 存在 by "being", the term used in contemporary philosophical language to distinguish it from *existence jitsuzon*実存; but in fact, *sonzai* connotes a *there*, *zai* 在. Unlike the Castilian *ser*, *sonzai* expresses a sort of *estar*

which would be a keeping-oneself-there-in-an-identity at the same time. In short, a being-existence. *Zai* 在 is a commonly used locative in modern Chinese, like the English equivalent "at, on"; for example in *zai jia*, "at home", *zai zher*, "here", *zai zhuozi shang*, "on the table", etc. It is similar to the *da* of *Dasein*, the German word meaning existence which Heidegger turned into a special concept, sometimes translated by "being there" but often left in the German "*Dasein*" by philosophers. Such a comparison goes beyond the formal level, as the following examples will show. In Japanese *kun* (semantic) reading both *son* 存 and *zai* 在 can be read as *aru*, a word which dictionaries give as the equivalent of "to be" but which in reality would be translated more faithfully as "being there" which presupposes a "there". *Hon ga aru*, means "there is a book" (e.g. on the table). The particle *ga* indicates the relation between the subject (*hon*) and the predicating verb (*aru*), as is *inu ga hoeru*, "a dog barks". However if you want to say "it is a book" (*hon de aru*), you somehow have to empty *aru* of the notion of space by delegating it to the particle *de*, which is locative (for example in *heya de neru*, "sleeping in the room"). With this delegation, the principle of which is totally alien to the copula "being" in English, *aru* can then become the equivalent of this copula; for example in *Mari wa kirei de aru*, "Mari is pretty". *Aru* keeps its spatial character and means "being somewhere, there is", except in that particular construct, which moreover functions in fact as if the subject (Mari) were *localized* in the predicate (*kirei*). In other words, in the category of pretty women *there is* Mari. These characteristics of the Japanese language have undoubtedly contributed to inspire Nishida Kitarô with the notion of "the logic of place" (*basho no ronri*), also called "logic of the predicate" (*jutsugo no ronri*), which overturned the logic of the Aristotelian subject (§ 11). One could compare it with the "primary metaphor" discovered by Lakoff and Johnson: *categories are containers* (*op. cit* in § 9, p. 51), while remembering that Aristotle calls *katêgoria* or *katêgoroumenon* what we would call the attribute or the predicate, from the verb *katêgorein*, short for *kata tinos agoreuein*, "to say something regarding something on the agora" i.e. in public.

16 Zeno negated movement since it is contradictory: both being and non-being in the same place. This Sophism is not as aberrant as it seems because there is no movement as such, only with respect to something. Cf. Galileo: "Movement is like nothing."

17 My translation is taken from Yamauchi himself (*op. cit.*, p. 3): 考えるということと、そういう思惟が存在することと同一である.

18 PARMÉNIDE, *Le poème*. New translation by Arnaud VILLANI, in collaboration with Pierre HOLZERNY, Paris, Hermann, 2011, p. 108. For fragment VIII, 34–36, Villani gives on p. 107 the following translation: "Thinking and that according to which one thinks, is the same thing. For without a being allowing him to express himself, you will not find thinking."

19 We follow here Jean-Baptiste de PANAFIEU, *Humanimal, notre zoo intérieur* (Humanimal, our inner zoo), Saint-Herblain, Gulf Stream, 2010.

20 Yamauchi does not seem to have been acquainted with works on logic sparked by quantum mechanics in the West such as those by Lupasco (1900–1988) with his dynamic logic of the contradiction, based on the principle of the *included* middle; see Basarab NICOLESCU, *Qu'est-ce que la réalité? Réflexions autour de l'œuvre de Stéphane Lupasco* (What is reality? Reflections on the work of Stéphane Lupasco), Montréal, Liber, 2009.

21 世界的思想体系の一翼, p. 67.

22 Precisely because this concerns your existence, as it does Hanako's as she is herself and not you, the same culture can paradoxically, it seems, emphasize

omoiyari 思い遣り (consideration, attention, compassion), in other words the fact of putting yourself in Hanako's place. This is not contradictory; on the contrary it is proof that you share the same medial body, an interlink (*aidagara* 間柄, § 20) where Hanako's existence and yours, growing together, have to take each other into account!

23 Pierre-Sylvain FILLIOZAT, *Le sanskrit* (Sanskrit), Paris, PUF, 2010 (1992), p. 17.

24 Japanese poetry has the same mind-set which does not separate *waka* (Japanese poetry) from other life manifestations, as is shown in Ki no Tsurayuki's famous introduction to *The Anthology of Ancient and Modern Poems* (*Kokin waka-shû*, compiled around 905): "Listening to the warbler singing among the flowers or the frog crouching in the water, makes it clear that there is no living being that doesn't sing his song / compose a poem." (*iki toshite ikeru mono izure ka uta wo yomazarikeru)* (translation Jacqueline PIGEOT, *Questions de poétique japonaise* (Questions regarding Japanese poetry), Paris, PUF, 1997, p. 9). The *ka* 歌 of *waka* 和歌 can also be read as *uta* and means both song or poem.

25 NĀGĀRJUNA, *Stances du milieu par excellence* (Fundamental verses of the middle way). Translated from the original Sanskrit, introduced and annotated by Guy Bugault, Paris, Gallimard, 2002, p. 61.

26 For the Sanskrit text, I use the Teramoto edition in five languages (Sanskrit, Tibetan, Chinese, Japanese, German) *Chûron* 中論 (Treaty of the middle), Tokyo, 1977 (1937), p. 40, who says here *ganatā* instead of *gantā*; however, the form *gantā* is used more often. The word has the same root as the German *gehen* or English *go*; hence the usual translations are *der Geher geht nicht*, and *a goer does not go*. See above, § 9.

27 There is no complement (C) here, but it does not affect the structure, which remains open to having one.

28 There is obviously something dia-bolical about the logos! However, let us not repudiate what has made modernity! In Asia, Taoism, just like Buddhism, preferred com-prehension, as shown by this passage in the *Zhuangzi* (*Qiwulun*, II, 1): "Vast knowledge embraces, small knowledge separates" (*da zhi xianxian, xiao zhi jianjian* 大知閑閑、小知閒閒).

29 *Op. cit* in § 24, p. 32.

30 VILLANI, *op. cit* in § 31, back cover and pp. 65–66.

31 I will refer here to a few passages from my article The *chôra* in Plato, cited in § 6. I use the translation of Plato's text from Albert Rivaud's edition, Paris, Les Belles Lettres, 1985 (1925), with a few minor changes.

32 See Alexandre CHTCHEGLOV, *Polis et chôra. Cité et territoire dans le Pont-Euxin* (Polis and chora. City and territory in Pontus-Exinos), Annales littéraires de l'Université de Besançon, Paris, Les Belles Lettres, 1992.

33 Chtcheglov, *op. cit.*, p. 17.

34 I must admit that Plato's text does not feature this verb explicitly (even though it is used elsewhere); however the idea of aporia itself is present, even in the superlative *aporôtatos*. We have an aporia where there is no (*a-*) passage (*poros*) for thought; hence *aporein*, to need, to have trouble, not knowing. Cf. the expression *aporô ho ti lexô* (Sophocles, *King Oedipus*, 486) "I do not know what to say (*legein*)"; before the *chôra*, the logos remains silent …

35 Alain BOUTOT, *Heidegger et Platon. Le problème du nihilisme* (Heidegger and Plato. The problem of nihilism), Paris, PUF, 1987, p. 224.

36 As we have seen in § 20, "regionality" (in German *Gegendheit*) is one of the two possible translations of the Japanese *fûdosei*, the other being *mediance*.

37 For a more balanced overview of Yamauchi's thought, see Kioka's article cited in § 30, The horizon of lemmic logic.

38 From the Indo-European root *MEDHYO*, "that is in the middle"; hence the archaic Greek *methyos*, and so *mesos*, "situated in the middle, medium" (and so *mésologie*), the Latin *medius* (same meaning), *medietas* (which gave me *mediance*), etc. the name of Madhya Pradesh (capital Bhopal), in India, the state situated between the Indo-Gangetic Plain and the north of the Deccan, precisely says the "State of the Middle". It is the synonym of *Zhongguo* 中国, the name adopted by China. Cf. also the title of the Chinese translation of the *Treatise of the Middle*: *Zhonglun* 中論 (Japanese *Chûron*).

39 Bugault's translation, *op. cit.* in § 32, p. 35.

40 P. 2 in Teramoto's edition, *op. cit.* in § 32.

41 Translation Bugault, p. 61. What Bugault translates by "which third might walk?" is in the Sanskrit text (Teramoto, p. 40) *kas tṛitiyo hi gacchati*. At the level of secular truths, i.e. in the domain of the logos (lemmas 1 and 2) the third (*tṛitiyo*) is indeed excluded.

42 Linguistics has used the term in a different sense. Here I simply give it the meaning of "to take (*lambanein*) together (*sum*)" both A and non-A, i.e. the fourth lemma.

43 This "in fact" has some relation to Bruno LATOUR's comments in *Nous n'avons jamais été modernes*, Paris, La Découverte, 1991. (*We have Never been Modern*, published in English in 1993.)

44 In the general sense (i.e. not only as biological species, but as a middle term between the singular and the universal) in which it is used by the other great representative (after Nishida) of the Kyoto School, TANABE Hajime (1885–1962), *Shu no ronri* 種の論理 (Logic of the species), Tokyo, Iwanami, 2010 (1934). In mesology, the species, or society, is the middle term concretely necessary between the universal (life or humanity) and the singular (the living individual, or the human individual). Tanabe, in the general spirit of the Kyoto School, equating species and the state, ended up like Nishida, philosophically legitimating ultra-nationalism. He repented after 1945.

45 Same *EI* root as the French *j'irai*, the Greek *ienai* (to go), the Latin *ire* (to go).

46 Roger-Pol DROIT, *Le silence du Bouddha, et autres questions indiennes* (The silence of Buddha, and other Indian questions), Paris, Hermann, 2010, p. 44.

47 Cf. itinerary, transition, initial …

48 Obviously there are others. Teramoto (*op. cit.* in § 32) for example, in the dedicatory stanza we have already cited, translates the same *pratītya samutpāda* by *innen* 因縁, that today specialists (such as Frédéric GIRARD, *Vocabulaire du bouddhisme japonais*, Genève, Droz, 2008, vol. I, p. 498) prefer as the translation of *hetupratyaya*, for "causation. Causes and conditions". For Yamauchi, this seems to go with the co-awaiting (*sôdai*) and really becomes co-originating.

49 相待 (Chinese *xiangdai*, Japanese *sôdai*) is the translation given by Chinese Buddhism of the Sanskrit *apekṣa* (the idea of taking into account) by borrowing the term from *Zhuangzi*, who utilizes it in the *Qiwulun* 斉物論 ("Of the flattening that makes things equivalent" according to the translation given by Anne CHENG, *Histoire de la pensée chinoise* (History of Chinese thought), Paris, Seuil, 1997, p. 111, the term appears in this passage II, 34): *hua sheng zhi xiang dai, ruo qi bu xiang dai* 化聲之相待、若其不相待, "discursive oppositions are only the absence of opposition. In other words, oppositions feed one another in the absence of opposition". Beneath any separation introduced by language there is a common ground that the lemmic, founded on the tetralemma, can grasp, not a *logic* founded on the principle of the excluded middle. This idea is close to what we have seen in the paradox of the walker who does not walk. It is the same radical critique of the pretensions of the logos: if for the West, "In the beginning was the verb (*ho Logos*)", for the East "the Dao that can say itself is

not the eternal Dao" (*Dao ke dao fei chang Dao* 道可道非常道) is the first sentence of the *Laozi*.

50 相対するものについてそのいずれをも否定するものである。

51 生滅について生を否定し滅を否定するものであつて、単に生滅を否定せんとするものではない。

52 Of course one recognizes here Zeno's Sophism.

53 These terms have obviously been formed by playing with the parallel between "stopping correspondence" (*zettai* 絶対, i.e. the absolute and "stopping attention" (*zetsudai* 絶待, i.e. abtention). To translate them I use the suffix *ab* (marking the distancing), the gap, the separation) instead of "cutting" (*zessuru* 絶する). Cut off (*ab*) from any correspondence, the absolute is alone (*solus*); and abtention means to be cut off from any attention, from any tension towards (*ab-tendere*, the virtual opposite of *attendere: ad-tendere*, which became *attention* and *awaiting or expectation*).

54 絶対は相対の否定であるが、相待の否定は空でなければならぬ。

55 This "tension towards" is close to *qu* 趣 (Japanese *omomuki*) which we have seen in connection with Zong Bing in § 14.

56 *Vocabulaire ..., op. cit.*, vol. I, p. 212.

57 *Ornament of the Precepts of the Grand Vehicle*, one of the foundational texts of Great Vehicle Buddhism, transmitted to Asanga the Bhudda Maitreya; French translation by Sylvain Lévi (1883–1935), *Exposé de la doctrine du Grand Véhicule selon le système yogacara*, Paris, Champion, 1911, rev. ed., Kyoto, Rinsen Books, 1983. English translation by Dharmachakra Translation Committee, *Ornament of the Great Vehicle Sutras: Maitreya's Mahayanasutralamkara with Commentaries by Khenpo Shenga and Ju Mipham*, Boulder, CO, Snow Lion, 2014.

58 *Wu ziti gu cheng, qian wei hou yizhi* 無自體故成、前為後依止。

59 I introduce the restriction of scale because today nobody can deny that the quantum non-separability of two particles that interacted in the past (see above § 27), which is a physical phenomenon, is somehow related to the symbolic. The hypothesis of a link between quantum physics and phenomena of consciousness has been proposed, among others, by Roger PENROSE, *Les ombres de l'esprit. À la recherche d'une science de la conscience* (Shades of the mind. In search of a science of consciousness), Paris, Interéditions, 1995.

60 A Planck length (10^{-33}cm, i.e. a millionth of a billionth of a billionth of a billionth of a centimeter) is the size below which the superimposition of states can appear by the non-reduction of the wave packet and coexistence of two space-time curvatures, like the famous example of Schrödinger's cat, both alive and dead. Above this length there is reduction of the wave packet, or de-coherence, and thus the cat can only be either alive or dead. By linking symbolicity with non-separability I propose the hypothesis – following in Penrose's shadow – that in the micro-tubules of the neurons, there remains some trace of the pre-de-coherence inherited from concrete experience, between what, after de-coherence, is split by the logos into A and non-A, thing and word, signified and signifier, etc.

61 *Zettai mujunteki jiko dôitsu* 絶対矛盾的自己同一. I have further comments on this example in *Écoumène, op. cit.* (see "Propos" above), pp. 56–57.

62 The organizer of the meeting (a student at the time) M. Sôkejima Shigeru, may have kept a recording which he hopes to be able to find some day. On this topic see the article by EGUCHI Kumi, Les souvenirs de M. Sôkejima, cited in § 30.

63 *Soku no ronri* 即の論理.

64 *Vocabulaire ..., op. cit.*, vol. II, 1476–1481, p. 1476.

65 As is expressed by the term *sokuji* 即時, *Logos et lemme* (Logos and lemma), p. 321.

66 As expressed by the term *risetsu* 離接, *op. cit.*, p. 312, "to separate–to join", which may be compared to "de-distancing", the Heideggerian (*Ent-fernung*), and to *reveal–conceal* we have already seen (§ 23) and will see again later (§ 36), as well as the hypothesis outlined above concerning Penrose, relating symbolicity to quantum inseparability.

PART III

Re-engaging

8 Nature makes sense for nature ... and beyond

§ 35. Sense, milieu, subjectity

The problematic of the milieu begins with the French word *sens*. From the perspective of mesology, as stated in this book's Preface, this word will be used in the French triple linked sense of *spatial-temporal direction, capacity of physical sensation* and *mental significance*, translated here into English as "direction", "sensation", and "meaning" depending on context. The first aspect relates to the ontological level of *the planet*: it is the physical direction in which things are going, in both space and time. The second is related to the *biosphere*: the capacity of living beings to feel and the corresponding organs. The third aspect is related to the *ecumene*: the ontological level where thanks to language things take on meaning elaborated and transmitted by our symbolic and technical systems. Just as the ecumene presupposes the biosphere which presupposes the planet (not the inverse), so the third layer presupposes the second, which presupposes the first and not the other way around.[1]

Mesology takes on the task of establishing the concrete links between these three levels of *sens* that are generally separated in other disciplines. Hence the neologism *mediance*, initially defined as "a milieu's *sens*" in the French triple acceptation.[2]

Correlatively, we generally differentiate between these three meanings. This is especially the case with first, *spatial-temporal direction*, distinguished from the other two, *capacity of physical sensation* and *mental significance*,[3] because that same word *sens* straddles the continental divide between the objective and the subjective domains.[4] A river flowing in a given direction is an objective fact that can be measured in various ways and relates to the laws of universal gravitation. Similarly, the wind's direction is linked to the Earth's rotation, etc. We are dealing here with the direction *in which* a phenomenon occurs. With life however, the direction appears *for* a given existent. For instance, gravitation and sunlight give trees the *choice* between growing upwards rather than downwards, a choice which a river does not have. This shows a certain degree of freedom that does not exist at the ontological level of the planet. To be sure, the degree of freedom of plants is not the same as that enjoyed by animals that can run up or down depending on circumstances, but it introduces an essential distinction between inanimate matter on the one hand and the living on the other hand. The living are capable of feeling and hence of directing their own behavior.

In other words, the living are capable of combining the first and second meanings of the word *sens*, while inanimate matter can only move in the sense of direction. It also means that the living are endowed with perception and proprioception, since they can both feel exterior things and correlatively define themselves one way or the other, and thus distinguish between self and non-self.[5] As opposed to mechanicism, this means that the living are endowed with subjectity – the fact of having a self – or at least a certain degree of it. Subjectity appears with life and continues developing to the point of producing the self-referential being of the *cogito*. The *cogito* however does not recognize the subjectity of other living beings, other than that of fellow humans. This cornering of subjectity, in other words dualism, is what leads to mechanicism.

To be sure, the sense in which Descartes understood mechanicism is outmoded today; but it still widely determines our ways of thinking, especially in biology. The philosopher of science Georges Canguilhem (1904–1995)[6] battled against it all his life. Mechanicism for instance wants to show that the upward growth of trees happens, among other *causes*, because auxin, through inverse phototropism, migrates in the cells of the stalk in the shade and makes that side grow faster, thus impelling the stalk's positive phototropism. (The word auxin comes from the Greek *auxein*, to increase; auxin is a growth-stimulating phytohormone, with the chemical formula $C_{10}H_9NO_2$, indole-3-acetic acid). Nice mechanism! However, it says nothing about the *reasons* why the subjectity of the vegetable kingdom *invented* this way of relating to light and gravity, while the mineral world cannot. The plant world is capable of making such an invention precisely because of a certain history from which its subjectity emerged. I will get back to this point in § 41. Animals on the other hand enjoy light or shade wherever and whenever they and their species choose to do so.

Mechanicism's refusal to acknowledge the subjectity of the living means also that it denies the capacity to choose. Foreclosing the *why* of the phenomena mentioned above, it sees only the *how* of the mechanism to which it reduces them, in spite of the fact that life and subjectity are identical. Indeed in death this is no longer the case even for our most subjective fellow human. Challenging the heredity of acquired characteristics, August Weissmann (1834–1914) for instance cut the tails of 1,604 mice over 22 generations, and found out that the tails of the 23rd were still as long.[7] He recognized that he only proved the non-heredity of the mutilation, but not of the characteristics acquired by the animal's active experience, as in Lamarck's theory. This is where the problem lies. To treat living beings as if they were devoid of subjectity, that is as objects, is to miss their essential core, which is their life; this is exactly what is foreclosed by dualism, the foundation of mechanicism. Of course, the subjectity of Weismann's mice did not inspire them to revolt against their torturer, or to lengthen their tail instead of shortening it out of contrariness or in compensation. The subjectity of a lineage of mice, or *a fortiori* of a plant species, is indeed not of the same order as that of an individual person of the species *Homo sapiens*: it is less intense and especially less focalized. Both however are hugely different from the objectity of a stratigraphic series in geology or of a cadaver ready for dissection.

The latest scientific advances factually confirm – only partially, and almost regretfully – what mechanicism's adversaries had long since affirmed in principle. Foremost among them, Imanishi Kinji (1902–1992), who went so far as to grant subjectity (*shutaisei*) both to species and to "the entire biotic society" (*seibutsu zentai shakai*), and even to molecules in as far as they are capable of movement.[8] Leaving these extremes aside for now, we will get back to this problem in § 43. Nevertheless, today it is no longer out of the question to imagine perception or even proprioception of life starting at the cell level, without falling into animism or vitalism.

The "Science and technology" section of *Le Monde* of December 24–26, 2011 (pp. 4–5) reported on experiments showing that stem cells are able to change depending on the rigidity of the substratum on which they proliferate as if they had a sense of touch. They even reorganize the nucleus of their internal skeleton and become either neurons or osseous cells, etc. Where before only mechanotaxis or chemotaxis was seen (the living following in the track of molecules); today we speak of haptotaxis (from the Greek *haptein*, to touch and *taxis*, arrangement), of durotaxis and even of plithotaxis (from the Greek *plêthos*, multitude), that is, the capacity to cooperate in groups of cells.[9] This verges on decision making … But stop with this subjectity! The article in question had a perfectly Cartesian title, "Mechanism of the living", even as Manuel Théry (at the Commissariat à l'énergie atomique), speaking of the regulation between cell and substratum, confesses that "the harder the support, the harder the cell applies force, and the harder it applies force, the more it draws on this support which then becomes more rigid. However, the cell should then contract ad infinitum; what stops it? Nobody has an answer" (*ibid.*). He does not go so far as to ask the unthinkable: *why does the cell stop?* An object does not *stop*: somebody or something stops it. Without subjectity, there is neither reason nor motive, but only causes. A mechanism interprets nothing and does not act; it reacts and that is all.

This mechanistic conviction has dominated modernity but it is today quite eroded. While it has always been contested from various non-scientific perspectives, in the 20th century science itself provided the most irrefutable challenge in the works of Jakob von Uexküll (1864–1944). He opened the way to ethology and biosemiotics. In *Streifzüge durch die Umwelten von Tieren und Menschen* 1934 (*A Foray into the Worlds of Animals and Humans, with A Theory of Meaning*, translated by Joseph D. O'Neil, University of Minnesota Press, 2011), where he summarizes the essential contributions of his extensive research. The introduction begins as follows:

> Whoever wants to hold on to the conviction that animals are only machines, abandons all hope of getting a glimpse of their milieus [*ihre Umwelten*]. ... animals are thereby pinned down as pure objects [*reinen Objekten*]. ... One forgets that from the outset the most important element has been suppressed, namely the *subject* [*das* Subjekt] that uses these means, affects and perceives with them. ... Whoever still holds the view that our sensory organs serve perception and our motor organs serves our action, will see in animals not just a mechanical device but will also discover the *operator* [*den* Maschinisten], embodied in the organs just as we are in our body. Then he will no longer consider animals as simple objects, but as subjects [*als Subjekte*], whose essential activity consists in perceiving and acting.[10]

The rest of the book brings a wealth of proofs provided by scientific experiments supporting a thesis that is as staggering as was Copernicus' or Darwin's in their time: *a living being is endowed with subjectity*. As such, it interprets the environmental given (*Umgebung*) to make it into its milieu (*Umwelt*), specifically adapted to its species, according to whose terms it adapts creatively, in the virtuous circle of its own world. And so, in the same environment (*Umgebung*), the milieu (*Umwelt*) of one species is not the same as that of another species. Correlatively, an environment unlivable for most can be an optimal milieu for certain species, called extremophiles, discovered by many later advances in biology. For instance, you have the *Pyrolobus fumarii* comfortable in hyperthermal water (it still reproduces at 113 °C); or the *Thermococcus gammatolerans*, that is not only a thermophile but also can sustain strong radiation. This concerns even pluricellular organisms such as the worm *Alvinella pompejana*, which lives in more than 80 °C heat in hydrothermal chimneys.[11] And so the rule discovered by Uexküll continues to be confirmed: even when the environment is very bad, the milieu can be optimal for some given being;[12] there is *mutual adequacy*, a harmony between milieu and species. Each species is best adapted to its own milieu. To speak of "objective reality" in this case becomes an abstraction: in Uexküll's terms, experiments have proven "that the implied hypothesis [of behaviorism] of an animal entering into a relation with an object is false".[13] The relation it enters into, and which constitutes its reality, is with the *things* belonging to its milieu, not the universal objects of the environment as they exist for ecological science. In this case, mesology would say: it enters into a relation with S/P, not with S.

Here is a specific example. An electromagnetic radiation of $\lambda = 700$ nm (nanometers) is a universal physical given (S). In our species, *Homo sapiens*, this wavelength is perceived (interpreted) as the color red (S/P). In the *Bos taurus* (cow) species, this same wavelength is not perceived as a color. It lies outside the range of visible colors. For a bull, red does not exist, and he cannot therefore enter into a relation with red, as Uexküll would say. What excites him are the toreador's gesticulations, not the color of the *muleta*. Similarly, the human eye does not perceive the infrareds perceived by the serpent's eye; neither does it perceive the ultraviolet perceived by the butterfly's eye.[14] Etc. Moreover, the specification of things particular to the human species includes an additional specification particular to each culture: (S/P)/P′. Red for instance, is not the color of wedding dresses in France, but it is in Japan. It signals stop for the average driver, but for the Red Guards of the Cultural Revolution, it signified: forwards! Etc.

Concrete things as such that have not been abstracted into objects are never universal or neutral. Starting at the physiological level through some form of subjectity, they always acquire a specific meaning and value, what Uexküll calls *Ton* (shade) which he organizes in several categories: *Esston* (shading into eating), *Schutzton* (shading into self-defense), *Wohnton* (shading into inhabiting), etc. Indeed, this *Ton* is obviously related to Husserlian intentionality: it is the shade according to which, in a certain cosmophany, the object (S) becomes a thing (S/P) *for* a living being. From a mesological perspective this "shade" is none other than the "as", according to which the milieu's reality is established for and by the living being in question: the "as food" (*Esston*), the "as habitat" (*Wohnton*), etc.

It is noteworthy that this "as" is always combined with a verb: *essen* (to eat), *schutzen* (to defend), *wohnen* (to inhabit), etc., that is, with words that Japanese grammar calls "movement words", *dôshi* 動詞. There is indeed continuous action. Uexküll even underlines the active aspect by speaking of *Tönung*, shading; for example, *Esstönung*, shading-in-to-eat, etc. The action in question then is a process: in mesology that is trajection, the trajection of S (the *Umgebung*) into P (*Umwelt*). This is far removed from the notion of substance. Things in the milieu are not substantial objects subsisting in their in-itselfness. They are always in the process of becoming in their interaction with the subject. Reciprocally, the subject is also always in the process of becoming in interaction with things.

Similarly, this means that, in living milieus, changes are not always a result of causality, i.e. of a mechanical sequence, but also of motivation, i.e. expectation and intention. Subjects expect their milieu to be what it is, while the milieu is so because of that expectation. In other words, we are dealing here with the co-awaiting and co-suscitation discussed in § 34; this leads us in return to infer that such notions pertain to trajectivity and not just to objectivity and subjectivity. They are medial and mesological. In short, this is the S/P reality.

Milieu then is not simply a given (an *Umgebung*); it is constructed through its relation with the subject that is constructed in turn in its relation to the milieu. Hence the mutual fit between animal and milieu, revealed by Uexküll's scientific experiments. To be sure, in his *Cours de philosophie positive* (1830–1842), Auguste Comte had already considered a possible interrelation between a living being and what he was the first to call a *milieu*. He even wrote: "According to the universal law of the necessary equivalence between reaction and action, the surrounding system cannot modify an organism without the latter also exerting a corresponding influence."[15] This elicited the following commentary by François Jacob:

> For Auguste Comte [unlike Lamarck], the surrounding ambiants and circumstances become *the* milieu, which then acquires a different status. It represents the whole of the external variables to which living beings are subjected … in short, everything that can act on a living body. However, it is no longer a one-way effect. The organism and its milieu have an effect on each another. … The organism cannot be dissociated from its milieu. The whole is modified and transformed.[16]

For Claude Bernard, later on, the same interrelation exists between "the cosmic milieu" (or exterior) and "the organic milieu" (or interior).[17] However, for positivistic philosophy as for experimental medicine, this refers to what Uexküll calls *Umgebung*, and subjectity does not come into it. Uexküll calls the study of the relation, in which subjectity does play a role, the *Umweltlehre*; this corresponds to *mesology* in the sense I use it here and in which Watsuji uses *fûdogaku* 風土学 when speaking of human milieus. This mesology brought a "theory of meaning" (*Bedeutungslehre*) that cleared the way for biosemiotics. Jesper Hoffmeyer, who made it into a discipline in the 1990s, proclaims his debt to Uexküll's pioneering work.[18] He also emphasized that the *Bedeutungslehre*, which makes the living being into the interpreter of a certain object (the *Umgebung*) as a certain thing (the *Umwelt*) pertains to Peirce's triads, and not Saussure's dyads (signifier/signified). Moreover, some analogies can be found between the trajective chain (((S/P)/P')/P")/P‴ … and Peirce's semiosis, where the "interpretant", i.e., the sign, can be reinterpreted indefinitely. In his preface to the second translation (2010) of *Streifzüge*, Dominique Lestel speaks of the future of biosemiotics and cites the work of two Israeli biologists who showed "that it is impossible to account for bacterial behavior by relying on exclusively mechanistic models". He adds that "natural intelligence to which bacteria belong, is already based on meaning and is far removed from artificial intelligence which is still based on *information*".[19]

Information indeed belongs to the first stratum of *sens*, the planet's ontological level (the abstract object, the universal and quantifiable S). We have known for a long time that the living embodies the second stratum which qualifies the first in a certain sense, the S/P reality in a certain milieu. However, biosemiotics and ethology, in other words the *Umweltlehre*, suggest today that it manifests a third stratum of *sens*, i.e. meaning: (S/P)/P′, which tradition had made into the domain reserved for humans, the only possessor of language. Today this needs to be more nuanced and certainly goes beyond the mechanistic reduction of any meaning to the shifting of bytes, in other words to the first stratum of *sens*.

§ 36. The trajection of the physical to the semantic

How do we get from information to meaning? The process starts with the *but also* of a syllemma (§ 33): both a spatial-temporal process, which, depending on scale, involves either the evolution or the history of a particular milieu or the ecumene in general, *but also* a process outside of time and space, in the immediacy of the symbol which exists without being what it is.

– What do you mean, a "process" that is both evolutionary and immediate?

– From the mesological point of view, this is a strange process given its *but also*. It escapes from the logos and cannot be digitized, and so becomes relevant to the distinction between the living and the machine: this process *is a trajection*. When some existent takes into account a given S through the senses, through action, language and thought, an S/P reality is born which is transmissible as a sign (S/P)/P′ to other existents from the same medial body. This *signi-fication* ("making-into-a-sign") interprets and represents S/P as something (here the sign P′) and so on, giving rise to a trajective chain, which, from assumption to hypostasis, evolves in time and space on diverse scales, *but also* always immediately present in the phenomenon of life. The living is the I interpreter for which S is P in the trajective chain (((S/P)/P′)/P″)/P‴, where concretely S is in fact always already an S′ (or S″, etc.), and P always already a P′ (or P″, etc.) This means that *in S and P, I is always already pregnant*, in other words the concrete reality of S as well as P is always already permeated with I's life.

This is the principle of what I will call *the poetics of the Earth*, extending from the most primitive form of life to the most advanced human consciousness. It is a twofold principle: *the indefinitely evolving pregnancy of I in the trajective chain (((S/P)/P′)/P″)/P‴* ..., and by this very fact *the indefinitely evolving construction of I's subjectity*, in and through the milieu (S/P) it obtains from the environment (S), which it grasps as P. Needless to say this perspective is alien to dualism and its focus on the object, and to the resulting mechanicism.

The scales of the trajective chain's evolution are metrical in the sense that they can be measured in time and space; for instance, neither in time nor space is the evolution of the species *Homo sapiens* of the same scalar order as the history of the Thirty Years' War. However, these scales are also ontological, outside of time and space and thus related to the immediacy of the syllemma comprising the symbol. Adopting a Heideggerian image, we could say that in the course of history the trajective chain passed from the reality of the "stone" era (i.e. the planet) which is "without-world" (*weltlos*), into that of the "animal" era (i.e. the biosphere) which is "world-poor" (*weltarm*), and from there into the "human" era (i.e. the ecumene) that is "world-forming" (*weltbildend*). However, the without-world is always immediately present in the world-poor and the latter in the world of the world-former. In other words, the planet is always there below the biosphere, which is below the ecumene, and the three levels react to one another in an ongoing feedback loop.[20]

The trajective chain of evolution and history, starting with the initial given S, i.e., the planet, leads to the increasing stratification of more and more complex and diverse worlds (P), provided that the "world former" refrain from devastating all the other worlds in the sixth large extinction while committing suicide. This was not foreseeable when in 1929 Heidegger gave his course on "the question: what is world?" (*die Frage: was ist Welt?*).[21]

Heidegger relied largely on Uexküll's research to answer this question. Strangely enough, given that this inquiry came shortly after *Being and Time*, he did not bother with either history or evolution. What I call here a "trajective chain" is absent from his analysis even though he dwells on "the as" (*das als*)[22] which in today's reading surely suggests the mesological "as". To be sure, it is a different perspective because Heidegger deals with "the fundamental concepts of metaphysics" while mesology specifically wants to focus on "natural history and human history". We cannot enter here into a discussion of Heidegger's thesis itself, but only introduce a few analogies in passing, keeping future inquiries in reserve.[23] In the *Grundbegriffe*, the *als* (as) is understood as "a structural moment of manifestness" (*Strukturmoment der Offenbarkeit*) (p.416) of the existent as such, which Heidegger compares further on to the *qua* in Latin (*ens qua ens*), and then to the ᾗ in Greek (ὄν ᾗ ὄν), and finally to a "statement" (*Aussage*) such as "a is b".

Let us stick here to the similarity of this "statement" with the basic formula of trajection: *S as P* (in other words, *S is P*), and the comparison made before in § 34 with "the logic of *soku*" according to Yamauchi (S *soku* P, i.e. *S is/is not P*), while remembering that the idea of a predicative world comes from Nishida. It is tempting to compare Heidegger's statement in the *Grundbegriffe* that "*world* signifies the *manifestation of the existent as such in its totality*" (*Welt* bedeutet *die Offenbarkeit des Seienden als solchen im Ganzen*, p. 435), to the idea of *cosmophany* (the appearing of a world) that we have already encountered in the Preface of this book. Cosmophany dealt not only with human worldliness, but also with the world of the living in general, a *geocosmos*, which is the general world of all the particular worlds coexisting on Earth. In other words, the world is what Imanishi calls "the biotic society in its entirety".

From the mesological point of view, the obvious homology between Watsuji's statement and Uexküll's certainly implies that the human is not the only "world former", which can only be so on the basis of a deeper and more general world formation, that of life on Earth. To be sure, Heidegger derived from Uexküll the idea that the animal is "poor in world". Concerning the famous case of the tick, the latter states explicitly that:

> The whole richness of the world surrounding the tick [*die Zecke umgebende Welt*] shrinks [*schnürt zusammen*] and is transformed into a poor image [*ein ärmliches Gebilde*], mostly composed of only three perceptive

> signs [*Merkmalen*] and three actantial signs [*Wirkmalen*]: that is its milieu [*ihre Umwelt*]. The poverty [*Ärmlichkeit*] of its milieu nevertheless conditions the certainty of activity, and certainty is more important than richness.[24]

There is an illogical interference between the two concepts that Uexküll's entire thesis tries to distinguish: the objective environmental given (*Umgebung*) on the one hand, and the milieu (*Umwelt*) on the other. In other words between the quantitative and the qualitative, between information and meaning. Objectively speaking of course, the tick's world is less complex than Uexküll's human world; however as world (*Welt*), it is just as complete. From the tick's point of view, one could also say what Plato said from his own human point of view about the world (*kosmos*) in the *Timaeus*, that it is "very large, very good, very beautiful and perfect" (*megistos kai aristos kallistos te kai teleôtatos*) as we have seen in § 16. What the tick perceives (its world) concerning its own being, is just as perfect as what, respectively, a human brain perceives, i.e. a human world.

Starting with the objective environmental given, the process of qualification produces a world as such, and leads in both cases to a phenomenal loss of information, which in the human brain has been estimated at millions of bytes per second.[25] That is, *a world becomes world only by foreclosing what does not concern it*. Or, to imitate Uexküll's formula "worst environment, best milieu", i.e. *loss of information, gain of meaning*. This is the whole difference between a human brain and a computer, or between the living and the mechanical: the latter records everything but understands nothing.[26]

– Aren't you falling into relativism here?

– No, because from the mesological point of view, the worldliness (P) of every milieu (S/P) never suppresses the referent (S). Meaning does not replace information; it always needs information as raw material, elaborates it, metabolizing it creatively for nourishment. Meaning is/is not information. The relation between the two is more complex than simply the reduction of meaning to a quantity of information (S) which alone would make it possible to speak of "richness" or "poverty".

Correlatively, the characteristic of worlds (P) is that they cannot communicate with each other since they are more qualitative than quantitative. They foreclose each other. Relativism would go no further, and determine that the "tickishness" of the tick's world simply cannot be put in human terms. However, since we are dealing with milieu (S/P), we must admit that the tick's world is *at the same time*, and hence by syllemma, poor and not poor in information. The first point: its poverty is not problematic; it is an objective observation which corresponds to Uexküll's remark. As for the second point, it follows from the same principle for any worldliness: worlds presuppose an existent I, endowed with subjectity, that creatively interprets the given (S) to make it into the (S/P) reality of its own milieu. This worldliness is what among the living gives rise to the incommensurability of what among humans becomes fully symbolic, language in particular. It is neither quantitative nor measurable, because it is just the faculty of stating and experiencing that A is at the same time non-A (fourth lemma: the syllemma and especially the symbol).

This makes meaning irreducible to information, but that is not the only reason. The other reason is what creates the link between A and non-A: the ternary that introduces the symbolizing middle term, namely, the living which concretely interprets S as P and thus is in the position of I in the relation S–I–P (S is for P for I). It is not in-itself, in other words in the abstract, that S and P are identified, as we saw above. Concretely it is by and for I, and in as far as S and P are permeated by I's life (or, which amounts to the same thing, in as far as they are experienced by I's flesh). In themselves, S and P are never reducible to one another: *the worldliness of the S–I–P relation is necessary, in other words it requires the subjectity of I.* By the way, this underlines the fundamental aporia of all binary semiology: beyond the primary layer of a simple shift of information in time and space (see § 35), there can only be meaning in a ternary relation.

– So meaning is necessarily subjective? You are falling again into relativism!

– What distinguishes mesology from relativism, in other words from simple constructivism, is that the latter is content with juxtaposing worlds (P) that are all completely absolute (and hence incommunicable) while mesology believes that the reality of every milieu is the result of a trajective chain (((S/P)/P′)/P″)/P‴ ..., which at a given stage of history or evolution, diverged from another one starting from a common base (S/P hypostatized, i.e. S′ etc.) The ultimate basis of the recursion is no other than the planet Earth (the primitive S). This primitive common base, as well as all the other secondary common bases that precede a branching out first in evolution and then in history, makes it impossible to rely on simple relativism. The in-itself of this primitive or secondary S is not only inaccessible in principle because the very fact of grasping it means to grasp it as something (S/P); however, it is also inaccessible in practice because we can only go back in history through representation (S/P).

For the same reason, mesology obviously differs just as much from a reductionism that quite incoherently absolutizes S while claiming to possess it, making it into the be-all and end-all of knowledge.

Clearly, the trajective chains that engender reality derive from exactly the same principle as semiosis, namely the indefinite grasping of S as P, by an existent. To be sure, what Peirce understood as "interpretant" is not what I understand as "interpreter", but that is not the issue. The principle remains the same: i.e., the semiotic triad which in Peirce consists of the three terms "sign–interpretant–object" and for mesology the three terms "subject–interpreter–predicate". For the same reason, trajective chains are semiotic chains that creatively produce value and meaning, *but they also* create substance by trajective wedging; thus they are the essence of evolution and history (see Chapter 10).

By the way, mentioning value inevitably calls to mind the economy. It is well known that this discipline has done much research into the origin of value.[27] As we have seen in § 22, according to mesology, value comes from the trajective gap that, like the Heideggerian *Entfernung* (distancing), reveals–conceals the relation between S and P. Value is not in the in-itself of the (S) objects but in the reality of things (S/P). Before the scarcity or utility of a certain resource can be considered and *a fortiori* quantified, a specific given in the environment (S) *needs to be instituted as a resource* (S/P) by some history in some milieu. This is not objective, it is trajective; we will return to this point in § 40. So Marx was right, but only two-thirds right: value does not only arise from *social* (techno-symbolic) relations but also from *medial* (eco-techno-symbolic) relations. Correlatively, it not only proceeds from history but also initially from evolution.

For the same reason and contrary to the ecological postulate, it is not absurd to imagine indefinite growth on a finite planet. We are dealing not just with a history of the planet (S) or even that of the biosphere (S/P) but with an ecumene (S/P)/P′. Of course, this does not relieve us of the indispensable task of reducing the ecological footprint of our lifestyle and, to use Hervé Kempf's phrase, it should not stop us from aiming at "a drastic economy of material and energy resources" with as leitmotif "fewer goods, more relationships!"[28] This is the inverse of the MOT for here "growth" means "human development" rather than "increase in GDP". However, this means that value is not reducible to the Earth; it arises from the trajective gap (S/P) between the Earth (S) and our world (P) that in principle is indefinitely extensible like any semiosis. This is what makes it possible that cultivating human interrelation (§ 20, here "links") can be a source of growth even when reducing the environmental impact of civilization. To denounce as most ecologists do the principle of indefinite growth, means to confuse biosphere and ecumene, the environment and milieu, information and meaning.

For mesology on the other hand, meaning and value emerge trajectively from history, and on another scale from evolution, *but also* from the immediacy of the symbolic. However long and complex it may be, a trajective chain is nothing but the syllemma *S as P*. Let us now flesh out this principle.

§ 37. Mediance of the living

We briefly mentioned Imanishi Kinji before (§§ 9, 21, 35). An entomologist, primatologist, ecologist, anthropologist, explorer and great mountaineer (he led the first ascent on the Manaslu, 8,153 meters high), Imanishi (1902–1992) is the most famous Japanese naturalist of the 20th century. His life and work are inseparable from Kyoto where he was born and died, and he also studied and taught at the university of Kyoto.[29] He is considered a representative of the Kyoto School, *Kyoto gakuha*, of which Nishida was the leading figure. As a primatologist, he was ridiculed for a long time by western specialists for his "anthropomorphism". Today his theses have become paradigmatic, as Frans de Waal recognized in a book that caused a stir a dozen years ago.[30] In a nutshell, the Japanese school of primatology he led applied to monkeys methods that the West reserved for anthropology: the study of kinship systems, etc. However, de Waal underestimated the position's ontological core; for Imanishi, subjectity is in no way the prerogative of humans alone. We will return to this more fully.

One of Imanishi's most fruitful concepts is *ecospecies* (*sumiwake* 棲み分け), discovered in his early research in the 1920s and 1930s concerning the ecology of ephemeroptera in the nymphal stage in the torrents close to Kyoto. He translated this concept as "habitat segregation".[31] However, this literal translation (from *sumi-*: inhabit; *wakeru*: separate), while not wrong, is far removed from the richness of the original concept. One of Imanishi's disciples, Sano Toshiyuki, prefers to call it *lifestyle partitioning*. Here is his definition of *sumiwake* in *Iwanami seibutsugaku jiten* (Iwanami dictionary of biology): "Biologically, the phenomenon of *sumiwake* can be defined as follows: two or more species, with the same capacity to live in a given 'place', divide the place into exclusive habitats."[32]

This adds very little to Imanishi's own translation, i.e., "habitat segregation". There were several possibilities but I chose *écospécie* (*ecospecies*) because that term embraces both the idea of habitat (*eco*) and of speciation (*species*) resulting from the dynamics of a structural moment (a *médiance*) between a species and its habitat. The originality of the concept of *sumiwake* lies in the insertion of that structural moment, where *-wake* has the double meaning of separation of the habitats (i.e., segregation) and the division of the species (speciation) in the concrete link between being and its milieu.

Thanks to this concept and the data inspiring it, Imanishi is shielded from one of the criticisms directed at Uexküll, for instance by Dominique Lestel in the preface to *Milieu animal et milieu humain* (Animal milieu and human milieu):

> Uexküll describes … the *Umwelt* as some kind of self-contained sensorial Leibnizian monad, and he does not ask what kind of relations might be established between the *Umwelten* of the different species that live together. In other words, the question of the "ecology of *Umwelten*" of the different species that live together, is of no concern to the great German biologist.
>
> (pp. 19–20)

On the contrary, Imanishi is particularly interested in those relations, whose concreteness is at the origin of the notion of ecospecies. This concept can account for the intra-species differentiation of ephemera; in other words, it could address the problem of speciation, and of inter-species sociality. The coexistence, and even concrescence between different species within the same environment, i.e. different *Umwelten* within the same *Umgebung*, would then become a question of the ecospecies.

This is a real difference between Uexküll and Imanishi. The first focused on a hermeneutics appropriate to worlds (*Welten*) belonging to different species, through the particular meaning (*Bedeutung*) which each gives the environment. As we have seen in § 37, these worlds (P) as such cannot communicate. On the contrary, the ecospecies is based on the double relation of species among themselves and the environmental given they have in common. Even though Imanishi does not refer to Watsuji and does not use his concepts, he is in fact dealing with milieu (S/P) and with mediance. But it is mediance at the level of species, not of culture as in Watsuji. Coexisting in the same environment impels each of the different species to elaborate more finely a "concrete life milieu": *seikatsu no ba* 生活の場, which term is in fact very close to *fûdo* (human milieu) in Watsuji. Ultimately, the onset of speciation operates through co-suscitation between milieu and species. The notion of ecospecies leads to a theory of evolution of a medial (S/P) rather than an environmental order (S) because it links significance (P) to information (S), and thus hermeneutics to ecology.

For that same reason, ecospecies is different from the notion of niche, which belongs more to ecology, a science of objects, than to mesology. See for instance the definition (in the ecological sense) in the *American Heritage Dictionary*: "1. the function or position of an organism or population within an ecological community; 2. the particular area within a habitat occupied by an organism." Neither meaning nor subjectity nor worldliness is taken into account. They are functors in trophic chains, without any distinction between the environment and milieu. Natural history remains the history of objects. Imanishi could not use Watsuji's conceptual framework since the publication of *Fûdo* (1935) did not precede his work on the ecospecies of ephemera; furthermore the concept of *fûdosei* (mediance) concerns only human milieus.[33] Even though Imanishi only speaks of environment, like Watsuji and Uexküll he puts subjectity (*shutaisei*) at the center of his inquiry.

As we have seen in § 21, he frequently uses a powerful formula: "the subjectivation of the environment, the environmentalization of the subject" (*kankyô no shutaika, shutai no kankyôka*), which is actually the adaptation of the two terms together in the concrete unity of the ecospecies. *Shutaika* acquires here a meaning that is close to "mastery", which is derived from the original sense of the element *shu* 主: master. The organization of its milieu by a subject (in this case an animal species) becomes at the same time its emancipation from the raw environment, as the following passage shows:

> Ants use all sorts of things as resources, showing thus the richness of their lives' content and their world's [*sekai*] range. They build an underground habitat where they even raise aphids and grow mushrooms. One could conclude that, except for our own human societies, there are none that subjectivize their environment so much [*sono kankyô no shutaika*], or free themselves so much from it [*kankyô kara no dokuritsuka*].[34]

Clearly, Imanishi's vision is close to both Watsuji's and Uexküll's,[35] even though he only uses the term environment (i.e., the *Umgebung* as given). Unlike his predecessors, he does not conceptually distinguish it from milieu, which in the case of ants is "formalized" just as ours is humanized; this seems to lead to contradictory formulations, as seen above.

– Indeed! This subjectivation-emancipation, where the environment non-A becomes the subject A, which is thus liberated from non-A. *What is up with that?*

– We need to take into account that Imanishi, inspired by reading Nishida, did not scorn the syllemma "A *soku* non-A" (A is/is not non-A). In Niwa's analysis of this method, one of the subchapters is called "environmentalization of the subject *soku* subjectivation of the environment".[36]

However, at the level of the living in general, this is the same as trajection which generates mediance in human milieus; this is the correlative process of "worlding of the body, the somatizing of the world" we saw in § 18. This opens up the following perspective: *mediance would be the condition for the evolution of species*. Then evolution would be a question of the relation to the *Umwelt* (the reality *by and for* the living as subject) rather than to the *Umgebung* (the reality *of* the living as an object). Since we know that *reality-by-and-for* overwhelmingly determines human history, it would be illogical if it did not somehow determine the history of the living as well, that is evolution … as long as we do not make the living into a simple machine! In response then to the *meaning* that the species gives its environment, evolution would not be limited to the mechanical effect of natural selection on mere random mutations.

In this respect, Imanishi never ceased to battle Darwinism. As early as *The World of the Living* he writes for instance (p. 153): "As the environmentalized subject gradually tries to subjectivize [*shutaika sen*] his environment, he becomes more and more environmentalized. This is the principle of adaptation."[37]

In the last chapter we will return to this relation between evolution and subjectity. For now let us keep to the mediance of the living. We have seen that the concept of mediance, defined by Watsuji as the structural moment of human existence, can be understood as the dynamic coupling between the animal body and the medial body in our species. Here the medial body is understood as the eco-techno-symbolic milieu, whose "being towards life" transcends our individual mortality. However, if we take away technique and symbol, can we still speak of mediance? Hence the double question: where do technique and symbol start? And are they really necessary for mediance?

Actually, conceiving of the duality of the living as partially trans-individual and transmortal is not something recent in biology. As Hervé le Guyader explained, there is a homology between Weismann's notions of *germen* and *sôma* and the notion of "the king's two bodies" in medieval jurists (§ 24):

> And thus, independently, the two bodies of the living, one going from generation to generation, and the other indefectibly linked to a generation are defined through mathematical reasoning in Mendel, and through a fundamentally evolutionary reasoning in Weismann. All future biology will be marked by this double constitution of the living organism. Without realizing it, biologists rediscovered theological spaces here at the heart of living matter.[38]

Genetics certainly has evolved since August Weismann (1834–1914), but the same duality can be found in the following lines by Werner Arber (Nobel prize in medicine in 1978):

> The simultaneous presence within the same genome of genes acting on behalf of individual organisms, called "household" genes, serving the individual in particular life conditions, and genes developing multicellular organisms, and evolutionary genes acting on behalf of the evolutionary development of populations, implies the existence of an intrinsic duality of the genetic organization. On the one hand numerous genes in the genome help living beings in accomplishing their own individual life. On the other hand, the evolutionary genes insure a regular expression of the living at the level of populations, according to changing life conditions.[39]

This example leads us to expect at the biological level, a homology with the animal body/medial body mediance found at the level of human milieus. Here the expression of evolutionary genes plays the role of the medial body which in human history insures the being-towards-life of the transmortals that we are (§ 24). Additionally, those same genes that insure the species' survival are equally part of the medial body of each of us. At play here is a double structural moment: an internal one (towards life) on the one hand, and on the other hand an external one (towards the world).

These views are clearly contrary to MOT nominalism. Just as the latter is antinomial to mediance at the cultural level, so it is at the biological level where it dominates since it belongs to the experimental sciences which generally don't bother with hermeneutics.[40] Here, the gene is homologous to the individual at the cultural level: a small MOT biologically justifying a larger one. The best advocate of that last position is Richard

Dawkins, with his famous thesis of the "selfish gene".[41] It focuses on the genes' natural selection, rather than on that of organisms and populations, in accordance with neo-Darwinian orthodoxy. The organism plays the role of the "survival machine" for the gene, while its intrinsic value consists in its aptitude to transmit the largest number of copies of its genes. In strategies that go beyond the individual organism itself, this could even lead to altruism, because in making sacrifices, it acts in the interest of its own genes.

We will not go into the details of this thesis. Dawkins, of course, uses "selfish" in a metaphorical sense. There is no intentionality whatsoever in the gene's evolutionary strategies, but only stochastic mechanisms. Let me just underline two points. The first is the marvelous concordance of the thesis with Mrs. Thatcher to claim that society does not exist (§ 12). Whatever constitutes the gene's milieu does not exist, except as the mechanical vector of its in-itself. The second point is that such unbridled substantialism leaves aside any meaning. There is obviously no question of reality-for (a subject) concerning the gene, but only of reality-of these objects such as genes, organisms, populations, etc. Clearly this is the opposite of a vision such as Imanishi's. There is only the subjectity of the *cogito*, the pure subject whose gene is the enantiomer: it is pure object. *Vivat dualismus!* …

§ 38. Identifying with another subjectity

… This dualism moreover wants to submit subjectity itself to the mechanism of the object. Such theses are uttered in the name of scientific objectivity, in other words, by abstracting the subject; thus eminently personal moral choices like altruism would simply express the mechanism of our genes, while culture would only express the mechanism of its memes (§ 22). The biological machine would then subsume the domain of the subject itself, whether it be individual or collective.

This self-devouring dualism, objectifying even the *cogito*, is nevertheless the inverse illustration of human mediance. Just as the congruence of the pure subject, the *cogito*, with its enantiomer, the pure object, unwittingly expresses the structural moment of human existence (§ 13), so does Dawkins' and Mrs. Thatcher's nominalism express, unbeknownst to them, the subjectity of the medial body (S/P, permeated by the subjectity of I, § 36). This view has determined individuals' thinking in this part of the world since the Middle Ages.

– You must admit that this has turned out well for them, but what does it have to do with biology? Aren't you being a constructivist here?

– No, there is indeed a structural homology between social nominalism ("there is no such thing as society") and neo-Darwinism, where evolution is reduced to a percentage of genes in a group, which is not a society but a "population". The MOT again. The substantial element is now a notch lower, from the organism to the gene, but already for Darwin, who died well before the discovery of the gene in 1909, natural selection concerned the substantial element of the organism, not of the species. Neo-Darwinism synthesized Darwin (1809–1882) and Mendel (1822–1884), who of course did not yet speak of genes but already thought that heredity was based on the combination of "elements".[42] The MOT again! It is in the same spirit of nominalism that population genetics, sociobiology and behavioral ecology ended up assuming a gene for each character and behavior.[43] This gene, without ties or place, like the *cogito*, mutates randomly and individually. To be sure, Darwin did not speak of genetic mutations, but his individualizing, disconnecting, and analytic vision is also derived from the same Aristotelian *topos*: the *ousia*, the primary essence, whose "hegemony [...] in theory and practice, [...] signifies the hegemony of the singular over the general."[44]

Whether we are dealing with the organism or the gene, we encounter that same MOT, a being without mediance or milieu, the in-itself whose *topos* does not need a *chôra* to be what it is, but simply material. Aristotle interpreted the *chôra* as matter, *hulê* (*Physics* 109b). As a good follower of Nishida, Imanishi of course had views that were at the opposite extreme. In a sense the ecospecies illustrates the first lines of *Basho*:

> To say that two objects have a mutual relation, form a system and maintain each other respectively, raises the question of what maintains the system itself, and where the system is established and where it finds itself. Whatever exists must be somewhere. … I will call that somewhere "place", inspired by what Plato in the *Timaeus* says of what must be called the thing that received the Ideas.[45]

Neither in Plato nor in Nishida is the milieu in question reduced to a material *topos*, even though it materially comprises *topoi*. It is ontological, or rather neantological in Nishida. He assimilates it to the predicate (*jutsugo* 述語) subsuming being (*u* 有) in a relative nothingness, a subsumption that expands into an absolute nothingness (*zettai mu* 絶対無). Let us leave aside the ontological and logical problems this raises.[46] The naturalist Imanishi on the other hand deals with concrete facts. He sees in the *basho* both the material *topos* where the ephemera live and the category (which in Aristotle is the predicate)[47] that subsumes those organisms, namely the species or the subspecies. His theory of the living expands this subsumption to "the whole biotic society" (§ 35), whose global link goes back to the origins of life, and more particularly to the Cambrian explosion.[48] None of these various levels of being is a primary essence. All are relative, all are real and, from the tiniest (the cell) to the most global (biotic society), they all share a certain degree of subjectity.

This gives rise to concepts such as ecospecies and speciety (*shushakai* 種社会: species as society). For Imanishi, there is a society whether organisms belong to species that are gregarious or solitary. The population is only an objectal abstraction; what matters in evolution is speciety, where organisms exist *for* each other, in as far as each of them and their society as a whole, are endowed with subjectity. For the same reason, biotic society includes inter-species sociality and so *existence-for*.

These views led Imanishi to plead for biogeography, as more open and concrete than ecology. Correlatively, he rejected Clements' theory of climax, which seemed too abstract (succession as simply temporal), and prefered Tansley's polyclimax, which allowed him to reconcile succession and ecospecies.[49] Imanishi's critique of Clements could be compared to Watsuji's critique of Heidegger in *Fûdo*, namely that the *Dasein*, too temporal and not sufficiently geographically concrete, is but an individual abstracted from his mediance (§ 24).

Imanishi's main interest is concreteness or rather concrescence. For him, focusing on the subjectity of the living means refusing to make it into an abstract object. He postulates an ontological link between the observer and the observed which arises from the origins of life and thus from biotic sociality. He develops what he calls "naturing science", *shizengaku* 自然学, to be distinguished from the natural sciences, *shizen kagaku* 自然科学, which reduce nature to an object.[50] This naturing science combines classic scientific methods with a hermeneutic phenomenological approach whose ontological presuppositions reject dualism and whose logical framework is actually based on the syllemma rather than on the logos. In this case the syllemma consists in the intuition according to which the observer as subject puts itself in the place of the observed.

Imanishi admits that this is easier with primates than with insects, etc. The fruitfulness of Japanese primatology has largely confirmed the value of that method; however, in the case of such things as insects, had not Uexküll's *Umweltlehre*, and especially his famous work on the tick, already proven that where the living is concerned, hermeneutical phenomenology could and should enrich science?

It is true that great naturalists have always been a bit hermeneutical even in the case of insects as with Jean-Henri Fabre for instance, or in the case of earthworms, as with Darwin.[51] However, even if we can see a common tendency, the divided identifications it implies are not as easily expressed when the mind and language have been programmed with the "I" of the *cogito* rather than inserted into the subject's circumstances (the "ambient") in the Japanese manner (§ 6). It is difficult for MOTs to imagine being an ant. And especially to say so … They would much rather be behaviorists!

However, with the physics of Heisenberg, even science has finally shown that "method can no longer be separated from its object" (§ 29). Today in biology one can claim to "think like a rat"; this is the title of a recent work by Viviane Despret who identifies herself as the "ethologist of ethologists"; as she clarifies:

> For a long time we believed we could discover the laws of learning by submitting a rat to labyrinth tests. To be sure, the rat will "learn" the route if it is rewarded. But which question does the rat really answer? What does the labyrinth mean to it? How does it interpret the reward? ... In fact, the animal would answer a question that is different from the one it was asked. ... Animals do not "react" to what we subject them to: they interpret a question and their answer translates their perspective on the situation. It is the latter that should be of interest ... "to think with" is both the issue and the condition of the research.[52]

This confirms the emergence of a new method in the natural sciences that provides a bridge to the social sciences: *a hermeneutical ethology*. This path has led to some spectacular projects but few have gone as far as Shaun Ellis, a British scholar who learned how to live with wolves, to the point of becoming the dominant male of a pack and teaching orphan cubs how to hunt![53]

This is indeed a case of inter-species sociality: the researcher enters into the communication system of his partners by treating them as subjects. Ellis for instance is careful to wear the same clothing when he meets his pack, to have his smell be familiar to the other wolves. To the "other wolves", yes indeed! To the nth degree, it is the principle of Imanishism as defined by Niwa Fumio: "It is the oriental way of knowing: by diving directly into nature, to try to seize it as it really is."[54]

An oriental way of knowing, undoubtedly, but drawing on the achievements of modern science (in this case ethology). In other words, a true overcoming of modernity (and not just a rejection), which means admitting and practicing what is foreclosed by the *cogito*: i.e., that human beings not only share a medial body with their fellow humans, but that they could share one with other living beings. They could identify with them, enter their world by learning some of their predicates, because they all share a more profound milieu which ultimately is the geo-cosmos.

The development of environmental philosophy has accompanied this paradigm shift in the last half-century, but on ontological bases which are generally alien to a mesology that takes into account the mediance of every living being as subject. This gives rise to moral considerations derived from the in-itself of objective ecosystems rather than the being-for of coexistence. Conceptually, if not in practice, the "deep ecology" of someone like Arne Naess, for instance, is as its name indicates an ecology that deals with the environment and not with milieu, where the human being is subsumed in the Great Self of nature without including in any way the problematic of subjectity. It is actually an ontology of the biosphere (and not of the geo-cosmos) abstracted from the ecumene, which in my view is fundamentally absurd even if it is strategically useful for the necessary reform of our civilization.[55]

§ 39. Nature understands itself in this way

In ecological holism the human being is simply considered as a functor of the biosphere. This is a reductive step backwards which forgets that life not only produces ecosystems; it engenders worlds. Among these worlds is ours, the condition of all our reality. Nature is therefore fundamentally ambivalent: it exists not only around us, but also in us, as the formative subjects of world.

As the "self-so", which is the meaning of the Taoist term *ziran* 自然, nature is both the universe and indefinitely, "the-always-to-be-born" (*natura*)[56] in every living being. It is born from itself to itself with each of us as we take it beyond itself trajectively. We need an ontology and a logic for this ambivalence, namely a mesology and a meso-logic, based on the syllemma, short of and beyond dilemmas such as either subject or object, either nature or culture, that have satisfied modernity for too long.

Poetry has shown the way, especially in East Asia:

衆鳥高飛尽	*zhong niao gao fei jin*	Above a flock of birds goes by
孤雲独去閒	*gu yun du qu xian*	A solitary cloud calmly floats away
相看両不厭	*xiang kan liang bu yan*	Gazing at each other tirelessly
只有敬亭山	*zhi you Jingtingshan*	There is only Mount Jingting

In this scene, the mountain and the poet (Li Bai, 701–762 CE) "look at each other" (*xiang kan* 相看), but their two beings merge in the mountain, that is in nature.[57] Another example can be read in the two verses of Tao Yuanming (365–427 CE) who finds authenticity when leaving the urban artifice and feels that his own nature is in harmony with nature itself. For him, to return to nature means to return to *his* own nature, in the unison between individual spontaneity and the natural course of things:[58]

久在樊籠裏	*jiu zai fanlong li*	Caged for a long time
復得返自然	*fu de fan ziran*	Again I can return to my own nature

The absence of articles in Chinese brings out the whole ambivalence of *ziran*, whose "self" is forever encompassing: *nature* and *my* nature are joined. This is both object and subject at the same time, trajectively united in the milieu by the structural moment of existence, i.e. mediance. While it is up to poetry to tell that story and art to show it – as for instance in Courbet's painting *The Origin of the World* – science needs to confirm it, instead of tracing things back three billion years to the time of the planet's simple mechanism.

This would be a naturing science *shizengaku* that would let nature emerge with us, instead of killing it by focusing on the object (§ 14).

Notes

1 This thesis is more developed in *Écoumène, op. cit.*above in the Preface, chap. 5 ("Sens"). However, in the trajective chain (((S/P)/P')/P'')/P''' ..., P (the sign, the world) tends to move further and further away from the initial referent. (S, the Earth), which gives rise to the metabasist illusion of the sign's autonomy (§ 27).
2 *Le Sauvage et l'artifice* (The savage and the artifice), *op. cit.* in the Preface, p. 165.
3 All three come from the same Indo-European root *SENT-*, which is already ambivalent: to go in some direction; to notice. As direction or orientation, *sens* is derived from the Germanic *sinnan*, to go towards, to try, to think of. The capacity for thinking and signifying, from the Latin *sensus*. There is in all this the idea of a tension towards something, that can be compared with the Chinese *qu* 趣, which we saw in connection with Zong Bing (§ 14), and with the Japanese *omomuki* 趣, one of the possible translations of *mediance*. What is at stake here is the structural moment of existence.
4 In the United States, the continental divide is the watershed between the water flowing to the Pacific Ocean and the water flowing to the Atlantic Ocean (or the Arctic in Alaska).
5 This happens in our body at every level, in particular that of the immune system protected by leucocytes (white blood cells). See Thomas PRADEU, *Les limites du soi. Immunologie et identité biologique* (The limits of self. Immunology and biological identity), Montréal, Vrin/Les Presses de l'Université de Montréal, 2009.
6 See *La Connaissance de la vie* (Knowledge of life), Paris, Vrin, 1965, trans. Stefanos GEROULANOS and Daniela GINSBURG, New York, Fordham UP, 2008.
7 IMANISHI Kinji, *Shutaisei no shinkaron* (Subjectity in evolution), Tokyo, Chûôkôronsha, 1980, p. 13.
8 Imanishi, *op. cit.*, pp. 209 and 211.
9 D.T. TAMBE et al., Collective cell guidance by cooperative intercellular forces, *Nature Materials*, 10(6), 2011, pp. 469–475.
10 pp. 21–22 in the 1965 edition, Hamburg, Rowohlt.
11 Wikipedia, "Extremophile", online. This substantial article has a good bibliography.
12 *Op. cit.*, p. 29 note 1: "*Optimale*, d. h. denkbare günstige *Umwelt* und *pessimale Umgebung* wird als allgemeine Regel gelten können." Philippe MULLER's translation, *Mondes animaux et monde humain* (Animal worlds and human world), Paris, Denoël, 1965: "An *optimal milieu* associated with *pessimal surroundings*, that is the general rule." Charles MARTIN-FREVILLE's translation, *Milieu animal et milieu humain* (Animal milieu and human milieu), Paris, Payot & Rivages, 2010: "As a general rule the worst environment could be the optimal milieu."
13 "die stillschweigende Voraussetzung, ein Tier könne jemals mit einem Gegenstand in Beziehung treten, falsch ist", *op. cit.* p. 105 ("the tacit assumption that an animal could enter into a relation with an object, is false").
14 Data gleaned in HIDAKA Toshitaka, *Dôbutsu to ningen no sekai ninshiki. Iryûjon nashi ni sekai wa mienai* (Cognition of the world in animals and humans. Without illusion, the world is invisible), Tokyo, Chikuma, 2003, *passim*.
15 In: "Cours de philosophie positive" ("Lecture on positive philosophy"); *Œuvres*, Vol. III, p. 235. Cited by François JACOB, *La logique du vivant. Une histoire de l'hérédité* (The logic of life. A history of heredity), Paris, Gallimard, 1970, p. 172.
16 *Ibid.*
17 Jacob, *op. cit.*, p. 205.
18 Hoffmeyer, *op. cit.* in § 27.
19 *Milieu animal et milieu humain*, *op. cit.*, p. 14.
20 According to Lakoff and Johnson (*op. cit.* § 9): the flesh interprets the data of consciousness in terms of flesh (therefore it occupies the position of a nominal

predicate, p. 58); however consciousness in turn interprets the data of the flesh in terms of consciousness, and so on.

21 Lecture in Martin HEIDEGGER, *Die Grundbegriffe der Metaphysik. Welt-Endlichkeit-Einsamkeit* (Fundamental concepts of metaphysics. World-finitude-loneliness), Frankfurt-am-Mainz, Klostermann, 1983, especially the 2nd part.

22 I only discovered this particular application of *als* thirty years after using *as* for geographical questions that at the time had no relation to metaphysics for me; particularly, the question of why nature began to be seen *as* landscape in the 4th century in China, and in the Renaissance in Europe. Discussed here in § 18, the theme of the *as* was developed in *Écoumène* (Ecumene) (particularly § 30: "The *as* of reality", but only in 2002 did I systematize it in the formula "S as P" which I develop in *La pensée paysagère* (2008) (*Thinking through Landscape*, trans. by Anne-Marie Feenberg-Dibon) and *Histoire de l'habitat idéal* (History of the ideal habitat) (2010). I discovered the *Grundbegriffe* ... even later, in 2012. This encounter between geography and ontology is the equivalent of what is called in pharmacology a "double blind test".

23 It deals with a close comparison with *Logos and Lemma*, written by Yamauchi before the 1983 German publication of *Fundamental Concepts of Metaphysics*, in particular concerning what Heidegger calls "The focus of metaphysics on the logos and logic: the basis of the non-original development (*nicht ursprüngliche Entfaltung*) of the problem of the world" (p. 418).

24 *Streifzüge* ..., *op. cit.* p. 29.

25 Presentation by Gilles ANSERMET at the colloquium Towards a naturalized hermeneutics?, Maison Suger, February 15, 2013, quoting Marcus E. RAICHLE, Two views of brain function, *Trends in Cognitive Sciences*, 14, 2010/4, pp. 180–190, p. 181:

> Thus, of the unlimited information available from the environment, only about 10^{10} bits/sec are deposited in the retina. Because of a limited number of axons in the optic nerves (approximately 1 million axons in each) only 6×10^6 bits/sec leave the retina and only 10^4 make it to layer IV of V1. These data clearly leave the impression that visual cortex receives an impoverished representation of the world, a subject of more than passing interest to those interested in the processing of visual information. Parenthetically, it should be noted that estimates of the bandwidth of conscious awareness itself (i.e. what we "see") are in the range of 100 bits/sec or less.

I thank G. Ansermet for making me aware of this article.

26 Let us point out in passing the mechanistic stupidity of those who deplore that we apparently only use 20% of our brain's faculties. The "unused 80%" (in this mechanical vision) are busy doing what distinguishes a human brain from a machine, namely living, and therefore creating meaning, for instance day-dreaming (§ 22), or dreaming. For these tasks that are essential to our humanity, the brain, with 2% of our body consumes 20% of its energy. What makes sense is the medial brain–computer coupling: by itself the computer has no meaning whatsoever and cannot do anything except rust. In mediance, the brain is the focal point; the computer is just a helper man adds to work in his place. This eventually might backfire and turn against this lazy King of Creation! Since Cro-Magnon Man, at least in terms of volume, our brain has been reduced by 15% (§ 21). We could reasonably ascribe this to technical progress, that is, to the deployment of our medial body reacting on our animal body.

27 See André ORLÉAN, *L'Empire de la valeur. Refonder l'économie* (The empire of value. Rebuilding the economy), Paris, Seuil, 2011. The author's thesis is

actually quite mesological: contrary to the dominant view, value is not an objective quantity (in other words, S). It always expresses a point of view (in other words, P) on behalf of some interests. I believe Orléan could say that value is S/P, or a human reality.

28 Hervé KEMPF, *Fin de l'Occident, naissance du monde* (The end of the West, the birth of the world), Paris, Seuil, 2013, pp. 96 and 106.

29 On Imanishi's life, see HONDA Yasuharu, *Hyôden. Imanishi Kinji* (Imanishi Kinji, a critical biography), Tokyo, Kôdansha, 2012. On his main views, ICHIKAWA Ryôichi, *Imanishi Kinji goroku. Shizen no fukken* (Quotes by Imanishi. The rehabilitation of nature), Tokyo, Shûfusha, 2008.

30 Frans DE WAAL, *The Ape and the Sushi Master. Cultural Reflections by a Primatologist*, London, Penguin, 2001. De Waal further developed his views in Silent invasion. Imanishi's primatology and cultural bias in science, *Animal Cognition* 4(6), 2003, pp. 293–299. He denounces not only the dogmatism ostracizing the theses of the Japanese School for a generation, but also the disregard of their origin after they had convinced the Anglo-Saxons themselves.

31 P. ex. p. 90 in *Seibutsu shakai no ronri* (The logic of living society), Tokyo, Heibonsha, 1994 (1948).

32 Cited by SANO Toshiyuki, The effect of culture on the development of scientific theory: Imanishi and Darwin, *Nara Women's University, Faculty of Human Life and Environment, Annual Report of the Graduate Division of Human Culture*, 19, 2003, p. 241.

33 Generally the term *fûdo* (milieu) is not used in the natural sciences. *Umwelt* in Uexküll's meaning, was translated in Japanese as *kansekai* 環世界 (surrounding world); and the usual sense in German – that is the objective environment, or *Umgebung* in Uexküll's terms – was translated as *kankyô* 環境, precisely what Watsuji distinguished from *fûdo*.

34 IMANISHI Kinji, *Seibutsu no sekai* (The world of the living), Tokyo, Kôdansha bunko, 1972 (1941), p. 95. This book was translated into French with the title *Le monde des êtres vivants. Une théorie écologique de l'évolution* (The world of living beings. An ecological theory of evolution), Marseille, Wildproject, 2011.

35 Even though *Seibutsu no sekai* does not refer to it, it is possible that Imanishi had in mind Uexküll's work when he wrote for example, p. 98: "Various living beings each find a different meaning in their environment" (それぞれの生物によって環境に見出される意義が違); more likely however, it comes from the emphasis he places throughout his work on the subjectity of the living.

36 *Shutai no kankyôka soku kankyô no shutaika* 主体の環境化即環境の主体化. NIWA Fumio, *Nihonteki shizenkan no hôhô. Imanishi seitaigaku no imi suru mono* (The method of the Japanese vision of nature. What Imanishi's ecology means), Tokyo, Nôsangyoson bunka kyôkai, 1993, p. 40.

37 環境化された主体はいよいよその環境を主体化せんとして、いよいよ環境化されてゆく。適応の原理はここにある。

38 Hervé LE GUYADER, *Penser l'évolution* (Thinking evolution), Paris, Imprimerie nationale, 2012, p. 153.

39 Werner ARBER, "Sagesse traditionnelle et savoir récemment acquis sur l'évolution biologique" (Traditional wisdom and recently acquired knowledge of biological evolution), in Jean STAUNE (ed.) *La science, l'homme et le monde. Les nouveaux enjeux* (Science, humans and the world. The new challenges), Paris, Presses de la Renaissance, 2008, p. 143.

40 On the power of nominalism in the life sciences, see Miguel BENASAYAG and Pierre-Henri GOUYON, *Fabriquer le vivant? Ce que nous apprennent les sciences de la vie pour penser les défis de notre époque* (Producing the living? What

the life sciences can teach us to think the challenges of our time), Paris, La Découverte, 2012.

41 Richard DAWKINS, *The Selfish Gene*, Oxford, Oxford University Press, 1976. Examples of the controversies caused by this thesis are featured in Kim STERELNY, *Dawkins vs. Gould. Survival of the Fittest*, Cambridge, Icon Books, 2007.

42 IKEDA Kiyohiko, *Shinkaron wo kakikaeru* (Rewriting the theory of evolution), Tokyo, Shinchôsha, 2011, p. 33.

43 Ikeda, *op. cit.*, p. 38.

44 ARISTOTLE, *op. cit.* au § 6, p. 42. (Of course these views are not those of Aristotle himself but of his commentators.)

45 NISHIDA Kitarô, *Nishida Kitarô zenshû* (Complete works of Nishida), vol. IV, Tokyo, Iwanami, 1966 (1927), pp. 208–209.

46 Nishida overlooks the fact that the *chôra* does not receive the ideas (Absolute Being) themselves but only their reflection (Relative Being). He absolutizes nothingness, not being. Logically, however, the subsumptions he talks about can only be an infinite regress, never ending in the negation of nothingness by itself, which is for him absolute nothingness and therefore, the source of being, which thus can only be relative. In Plato and Nishida, the absolutization of being or nothingness is a mystical leap that neither of them is aware of.

47 See above, § 31.

48 In the beginning of the primary era, the Cambrian, between 543 and 530 million years ago, life became suddenly and extraordinarily diversified at the same time that the Ediacarian fauna disappeared; that fauna had appeared one hundred million years before, and is now an unknown life form.

49 *Seibutsu shakai no ronri*, *op. cit.* in § 37, chap. IV; and Niwa, *op. cit.* in § 37, p. 79. Frederic Clements (1874–1945) was a pioneer of plant ecology. In 1916 he invented the concept of *climax*, the final and stable state of an ecological *succession* after the re-colonization by life of some area, notably in North America after deglaciation. The notion of *polyclimax* refined that of climax, by taking into account that, in the same climates, there can be several kinds of climaxes depending on local conditions (conditions of the soil, etc). On these questions, see Pascal ACOT, *Histoire de l'écologie* (History of ecology), Paris, PUF, 1988; and Frédéric BIORET, Roger ESTÈVE and Anthony STURBOIS, *Dictionnaire de la protection de la nature* (Dictionary of nature conservancy), Presses universitaires de Rennes, 2009.

50 IMANISHI Kinji, *Shizengaku no teishô* (For a naturing science), Tokyo, Kôdansha, 1984.

51 To which he dedicated his final book, *The Formation of Vegetable Mould through the Action of Worms, with Observations on their Habits*, 1881. See also Adam PHILLIPS, *La mort qui fait aimer la vie. Darwin et Freud*, Paris, Payot & Rivages, 2005 (*Darwin's Worms*, 2000), who writes on p. 77: "The type of work he [Darwin] carried out nevertheless testifies to his capacity of developing diverse and subtle sympathies (in psychoanalytical terms this would be called 'split identifications': to imagine being somehow part of groups that seem mutually exclusive)".

52 Viviane DESPRET, *Penser comme un rat* (Thinking like a rat), Versailles, Quæ, 2009, back cover and p. 48.

53 Wikipedia gives as Shaun Ellis' bibliography the following: *The Wolf Talk*, Rainbow, 2003; *Spirit of the Wolf Talk*, Parragon, 2006; *The Man who Lives with Wolves*, HarperCollins, 2009.

54 Niwa, *op. cit.* in § 37, p. 229: 自然にじかにのめり込むかたちで即自としての自然を把握しようとする東洋的認識様式.

55 I developed this critique of Arne Naess' main work in Ontology of ecosystems, or of human milieus?, *EspacesTemps.net*, June 15, 2010, http://espacestemps.net/document8138.html/ The absurdity here consists in expecting moral responsibility from humans while considering them as simple eco-systemic functors. More generally, see my *Être humains sur la Terre. Principes d'éthique de l'écoumène* (Being human on Earth. Principles of ethics of the ecumene), *op. cit.* in § 25.

56 In Latin, *natura* is the feminine of future participle of *nasci*, to be born: i.e., "the one to be born"; see also below, § 45.

57 An oronym meaning: Pavilion of respects. Poem by Li Bai (701–762), *Du zuo Jingtingshan* (Alone on Mount Jingting), taken from *Tang shi jingxuan* (Anthology of Tang poetry), Beijing, Jindun chubanshe, 2003, p. 138.

58 Extract from *Gui yuantian ju* (I return to live in the countryside), MATSUEDA Shigeo and WADA Takeshi (eds.), *Tô Enmei zenshû* (Complete works of Tao Yuanming), Tokyo, Iwanami Bunko, 1990, vol. I, p. 96. "Again" (*fu* 復) indicates that this is Tao Yuanming's second return to the fields (on this theme, see my *Histoire de l'habitat idéal* (History of the ideal habitat), § 36: "Hermitage"), and does so more specifically than Paul Jacob (TAO Yuanming, *Œuvres complètes* (Complete works), Gallimard, 1990), who for these two verses gives the following translation (p. 129): "For a long time I was in a cage / To the spontaneous I return again".

9 The contingency of life itself

§ 40. To bring into existence as

The principle of trajection is as follows: for some being (the interpreter, let us say I) an environmental given will exist as something (P) whose reality is the S/P relation, namely "S as P": this ternary relation (S–I–P) allows S to exist as P. Without this, S would remain a latent power, a potentiality waiting to be discovered – but it is therefore discovered as something summed up by four great predicative categories: *resources, constraints, risks and pleasures*. Examples of resources (P) for the cheetah (I) are: the antelope (S) as prey (P); for the oak leaf (I), sunlight (S) as energy for photosynthesis; for Einstein (I), *c* (S, the speed of light in a vacuum) as a constant on which to base the special theory of relativity (P). Examples of pleasures: for me (I) this veal or venison leg for lunch (P). Etc.

There is already such a symbolization in one of humanity's oldest texts: a famous passage in the *Odyssey* (X, 302–306), famous because it contains the first occurrence of *phusis* which later means "nature". Here Hermes, pulling a wild herb from the soil, teaches Odysseus the quality that will prevent him from being turned into a pig. Only the gods know this plant's name, *môlu*, but thanks to the industrious Hermes, it will exist for Odysseus as a *pharmakon*, a medicine, that is, a resource. So now the *môlu*, trajected into the human world through technique and symbol, will live on as an antidote, as shown in later texts.[1]

That is not all. Homer implies that the trajection of the *môlu* (S) as *pharmakon* (P), has a double meaning, remedy or poison, as reflected in its white flowers and black roots. In other words, its reality (S/P) depends on each case; it is contingent, historical, neither necessary nor natural. In short, it is trajective: only the gods know what the *môlu*'s (S) in-itself reality is; besides to bring it into our world by pulling it out of the earth (*ek gaiês*, 303) is difficult for mortal humans (*chalepon … andrasi ge thnêtoisi*, 305–306). The trajection of S as P that engenders reality is indeed the hard "struggle" that according to Heidegger, opposes the Earth to the world (§ 19).

As we have seen in Chapter 8, trajection is set in motion along with the biosphere, in the worlds of the living. It is the passage from the abstraction of the *Umgebung* to the concrescence of the *Umwelt*. This contingent history surely begins with life but probably even before life, in the physical world, or even in mathematical laws. What in celestial mechanics is called "the three body problem" (for example two planets and a star) shows mathematically the impossibility of predicting how orbits will evolve over time. The uncertainty, which is integral to the mathematical calculation of the ternary, "gives time its direction: an arrow aiming from the most orderly (the past) to the least orderly (the future)".[2] Concrete history where the number of bodies is virtually infinite, exponentially increases this uncertainty, before moderating it by the law of large numbers. Let us remember however that as soon as we have a ternary, the future becomes uncertain.

There is another approach which pertains to experimentation in quantum physics: what Niels Bohr in 1927 called the "complementarity" between the undulatory and corpuscular aspects of the same particle. In other words, the latter will appear either as a wave or as a particle, depending on the experimental device. Reality diverges here from logic: a sort of "interpreter" (I, the experiment, although purely objectal) the in-itself S of the particle exists either as A (a wave) or as non-A (a corpuscle). Concretely this particle is both A and non-A. Similarly, in quantum inseparability, the same particle can be both in an A place and in a non-A place. Experiments have confirmed this ever since. In the latest one, the team led by Nicolas Gisin in Geneva succeeded in "intricating", that is to make unitary, as one single object, not only two particles, but two macroscopic objects: a green crystal and a red one.[3] Mesology concludes that already at the physical level, reality is based on the syllemma (both A and non-A) and presupposes a ternary (A, non-A and I: the interpretative mechanism).

Experience – in other words history – is what makes S exist as one thing or another. Let us leave physics,[4] as our study is limited to re-engaging human history with natural history, the ecumene with the biosphere where the interpreter I is no longer at the ontological level of the planet, simply an object (an experimental device),[5] but a subject that has variable ontological levels. The subject's level of being is defined by the deployment of world (P): the greater the range of predicates (P, P′, P″ etc.), i.e. the longer the trajective chain, the greater the world formation (*Weltbildung*) will be, and the higher the ontological level. If there is no world, there is no subjectity; the more world, the more subject the subject will be. Schematically, one could define the worldless-ness (*Weltlosigkeit*) of stone as S, the world-poverty (*Weltärmlichkeit*) of the living as S/P and the properly human world formation of being as (S/P)/P′, thanks to language which supports the subject–predicate of selfhood, in other words consciousness (§8). This means that like the *cogito*, human beings create their own existence as such and may imagine that the predicate instantiating them frees them of any milieu. In fact they can only utter that predicate because they are alive and on Earth. They are at the end of the long trajective chain of a history and even before that of an evolution, and always in a correlatively evolving milieu.

In as far as S/P differs from a simple iteration of S, trajection is creative; it is a poetics in the sense of *poïêsis* (creation, fabrication, conception), where S *ek-sists* outside its identity to become something that is no longer simply S. For instance, in the case of chlorophyll assimilation, luminous energy (S) becomes an oak (S/P); or lunch meat is metabolized and becomes me: (S/P)/P′. On a greater level, in about four billion years, the Earth's poetics has engendered the world we know. In all this, we go from the universal to the particular, not the other way around.

It is not important to analyze the different mechanisms in themselves (there would be an infinite number), but on the contrary to find their common principle – trajection as an evolutionary and historical principle. This avoids creating a dualistic gap between human subjectity and other subjectities by relegating the latter to the mechanical status of objects. Science itself has been filling this gap, not only concerning the animal kingdom but even the realm of plants. There is today a *plant ethology*, featured recently on the cover page of *Science & Vie* (1146, March 2013), enticingly reporting on what is nevertheless very academic research: "Plants. They are intelligent!" For instance on page 51: "who would have thought! Plants have hearing, they know how to move and communicate, they have family loyalty and even have memory! In short: they are 'intelligent beings". On page 52:

> The fact they are fixed is the reason for the extreme complexity of plants: they are obliged to find metabolic responses to the dangers that threaten them. This explains why plants have more genes than animals. ... Rice has twice as many of them as humans!

However, I am not one of those whose "strange furor to be a monkey" is denounced by Alain Prochiantz.[6] Like him, I am not fooled by the "spectacular number 1.93% supposedly quantifying the genetic difference between humans and chimpanzees".[7] There is indeed a *qualitative difference* responsible for e.g. our disproportionate brain according to the norm of our closest cousins: "we have an 'excess' of 900 cubic centimeters. Given the normal 1400, that is quite a lot".[8] Compared to the 1.93% of genomic difference, it is not trivial to have a three-quarters larger brain, and not just any kind, but one that includes the cortical regions of language for instance, practically absent in the chimpanzee. But to say like Prochiantz that "animals are not subjects",[9] is to go no further than the "I" of the *cogito* and its world of objects. Mesology prefers to follow Uexküll (§ 35) and the teaching of the wind chimes (§ 4): subjectity is in no way the preserve of the "I". Even if human subjectity is in its present state the result of a spectacular acceleration since humans diverged from monkeys, it has been built, layer after layer, by a trajective chain going back 3.8 billion years, and whose base involves the totality of living species.

§ 41. Chance, contingency, necessity

With life appears the interpreter that not only looks for what surrounds it but also gives it meaning. It not only perceives its surroundings, but also acts on them and thus reproduce their existence as its own milieu. This brings about contingency which is neither chance nor necessity. The difference between chance and necessity is not resolved by an alternative that only a stochastic calculation can overcome, as in the case of quantum probability. Chance can be anything, anytime and anywhere: necessity is the repetition of the same always and everywhere. Contingency is in between these two: things could be different from what they are, but they are concretely what they are because of some history and some milieu that have meaning *for and by some interpreter*, a subject, whether individual (for instance the "I" of the *cogito*) or collective (for example the entire biotic society).

In this case dualism consists in cutting this reality into two irreconcilable domains: on one side you have the objective, a feature of either chance or necessity; on the other side the subjective which possesses a *sui generis* dimension, i.e., the arbitrary. This arbitrary accounts for the relation between subject and object (for instance, Mrs. Cogitans-Loquens decided that "this object shall be called 'table'") while the relation between the object and the subject is a result of either chance or necessity, which are objective categories (for instance "the falling rock crushed Mrs. Cogitans-Loquens who was passing there accidentally, but in accordance with the laws of falling bodies"). However, given the number of subjects, the effects of their arbitrariness in the end come down either to the chance of small numbers or the necessity of large numbers, i.e. to the laws of statistics.

This is the mental framework that produced Jacques Monod's famous essay, *Chance and Necessity*,[10] in which he uses for example the concept of "teleonomy" to account scientifically for something that the prescientific mentality – vitalism and animism[11] – took to be finality.

Rejecting the alternative between chance and necessity, at least when it concerns life, mesology is betting on a third and different genre, contingency. Two examples, a poem and a calculation, will summarize this option.

Caminante, no hay camino	You are walking, but there is no path
Todo pasa y todo queda,	Everything passes and everything stays,
pero lo nuestro es pasar	However we must go on
pasar haciendo caminos	Go on while tracing paths
caminos sobre el mar …	Paths on the sea …
Caminante, son tus huellas,	You who walks, these are the traces of
el camino y nada más;	Your path and nothing more;
caminante, no hay camino,	You who walks, there is no path,
se hace camino al andar.	The path is traced by walking.

Al andar se hace camino
y al volver la vista atrás
se ve la senda que nunca
se ha de volver a pisar …[12]

You trace your path by walking
And when you turn around
You can see the path
That you will never have to tread again …

And here is the calculation:

> The notion that evolution is due exclusively to successive micro-events, to mutations each happening randomly, is contradicted by time and arithmetic. To extract from a roulette, separately, one subunit after another, each of the hundred protein chains that make up the body of a mammal, time is needed that far exceeds the duration allocated to the solar system.[13]

One generation later this was clarified as follows:

> The molecules responsible for practically all biological functions, the enzymes, are proteins, that is, chains of at least a hundred amino acids set end to end. Natural proteins use twenty or so amino acids. There are at least 10^{130} possible different proteins. Let us suppose that each atom of the observable universe (there are around 10^{80}) is a computer, and that each enumerates one thousand billion combinations per second: this would exceed present computer capabilities. One would need one thousand and twenty-one times the age of the universe to get to the end of the list! Only one minute fraction of those possibilities is compatible with life as we know it. The universe is much too young for the process to have occurred only through some mechanism of systematically random attempts exploring all possibilities.[14]

Since mesology turns a deaf ear to the braying of creationism and to its scientistic mask, "intelligent design", what does it have to say about such a calculation? It means exactly what the preceding poem says: while life has no aim except to live, yet it has a meaning which results not from following a preordained path, but which is defined progressively by its own evolution. The latter gradually forecloses the possibilities which do not belong to the worlds it chooses for itself, by and for itself – *ziran* 自然, "by itself thus", as Taoism would say. In other words, the subjectity elaborated as life evolves, creates a meaning one can make out in retrospect (as do paleontologists and historians), but where it leads no one knows. This is the case at every level of subjectity, from the unconscious hyper-collective of biotic society to individual consciousness.

This meaning is not inscribed anywhere, but emerges written in the evolution and history of any and every milieu. We interpret reality as we go along on the journey. Meaning is not arbitrary, but contingent; while diverse possibilities open up all the time (there is no pure necessity), other possibilities are closed gradually (there is no pure chance), meaning thus continuing more and more in its own singular tracks. Cuvier already understood this when he invented in 1805 the term *embranchment*. He distinguished four types: Articulata, Mollusca, Radiata and Vertebrata.[15] The terminology has evolved, but the idea remains the same: once committed to one of the paths, life cannot take another, while in the beginning it could create any one. Obviously it did not refrain from doing just that, as shown by the extraordinary enigmatic fauna of the Ediacaran period (635–541 million years ago). Chance and necessity are atemporal abstractions pertaining to the view from nowhere that science focuses on the environment (*Umgebung*). However in concrete milieus (*Umwelten*) contingency reigns, or even what Yamauchi (§ 34) calls co-awaiting, because there is meaning for all those involved.

Given this perspective, I would like to follow Josef Reichholf when he argues, against the neo-Darwinian orthodoxy, that natural selection can only stabilize, but not innovate. Rejecting gradualism even more than the theory of "punctuated equilibrium",[16] he believes that "evolution emerges in disequilibrium".[17] Correlatively, he gives a central role to an organism's metabolic link to its milieu. New functions such as flight, cannot appear through a succession of small random adaptations to compensate for some insufficiency, but on the contrary require a sudden functional diversion of surplus potential, and the exploitation of new life possibilities offered by that potential.[18] "This is no longer about blind chance as the starting point, but of exploiting concrete possibilities or necessities. Chance is marginalized."[19]

Reichholf's thesis seems to agree with what ecology teaches us: an ecosystem in good shape has reserves that make it even more sustainable and resilient. Let us quote Michel Griffon on the ideal agriculture that would mimic natural ecosystems instead of fighting against them:

> The concept of ecological intensity was created to deal with the notion of sustainability: a system's sustainability is maintained when the functional fluxes of ecosystems are increased so that they somehow evolve towards a "synthetic climax" corresponding to the maximum intensity of functional fluxes [carbon, nitrogen ...]. That desired state must be as resilient as possible, in other words capable of resisting probable variations in the environment by creating functional stocks [carbon, water, soil nutriments ...]. To recover resilience capacities, the developed ecosystem must evolve from a tendency to simplify and standardize, which means degradation, towards a tendency to greater diversity and complexity ... which means an "aggradation", according to the term used by Professor Lemieux.[20]

I would like to show a parallel between this so-called "aggradation" and the surplus mentioned by Reichholf. The ideal situation that would provide life with more resilience: isn't that exactly the inverse of what mechanicism and its friend, capitalism, are aiming at? They idealize the elimination of surplus supplies, abolition of stock and the "just in time", in other words the reduction of any process to its simplest iteration, while the resulting weakening of systems is compensated by increasing negative externalities such as the discharge of the costs onto the surrounding milieu, which they degrade instead of aggrading. According to the same hypothesis, the neo-Darwinian vision, privileging competition and struggle for life over cooperation and co-originating, also reflects the dominant mechanicism and capitalism. To be sure, the parallel in social Darwinism between biology and society has long since been established,[21] but I would emphatically reject this ideology by showing the close link between evolution and resilience. For life to be able to invent new species through resilience, shouldn't surplus possibilities be available?

Surplus, but of what and to what end? Reichholf does not speak of subjectity, or *what* exactly exploits these "possibilities", in other words, what makes them exist as resources, if not some interpreter? What is a sudden "functional diversion" if not a choice by some interpreter, of a new *as* (P) concerning an S that was already there as a surplus for something else? All this is quite trajective … and suggests the following hypothesis: the reading (P) and hence the expression (S/P) of a more or less identical genome (S) may vary (§ 42). From this viewpoint I would like to consider a concept close to Reichholf's thesis: *exaptation*, introduced by Gould and Vrba in 1982.[22] Different from adaptation, exaptation is not an *aptation* to the present conditions of the environment, but a functional change.[23] For example, the feathers that ornamented the forelegs of some flightless dinosaurs, and whose function was to lower their temperature, were exapted to flight.[24]

For mesology, all of evolution must be considered triadically (in the ternary relation S–I–P, § 40), in terms of milieus, and not solely in terms of the environment; in those milieus the phenomena occurred concretely and contingently, neither by chance nor by necessity. They made sense for and by those involved. To think only in terms of the environment leads neo-Darwinism to paralogisms such as the hypothesis of the red queen,[25] according to which some species are condemned to evolve in order to remain in synch with other species that are their prey or predators, and with which they therefore co-evolve. This is of course coherent; however, why would this mechanical illustration of the universal principle of Darwinism, the struggle for life, not be valid for other species that barely evolve?[26] Simple casino game of chance? Isn't it rather that each species gives its environment a meaning, makes its own milieu, within which, trajectively, it either does or does not evolve?

§ 42. Life's trajectivity

We have seen that for mesology, what occurs between the physical and the phenomenal (§ 29) or between physics and semantics (§ 36), is the same as trajection (S/P)/P′, where S/P becomes S′ with respect to P′. Since the agent of this trajection is life itself, we must presuppose a strong link, or even the identity between those two notions. *A contrario*, death stops trajection: the organism no longer has a milieu (S/P), and returns to the environment (S). It becomes world-less, *weltlos*, like stones, which in turn of course benefits other organisms. This invites us to compare trajection to metabolism, which keeps the organism alive, and which stops death (at least for the one involved). In fact, if to metabolize means to transform a substance into metabolites which makes growth and the multiplication of cells possible, a chemical operation taking place thanks to "some class of proteins, enzymes functioning as specific catalysts",[27] then this could be represented by the same formula as a trajection, where the initial substance S becomes S′, the metabolite.

– Aren't you spinning out the metaphor a bit too much?

– What is a metaphor? It is carrying (*pherein*) further (*meta*) the meaning of a word. For example, comparing the sun to a ball and the Earth to an orange turning around a ball. Of course, the sun is not a ball; but it becomes one in this metaphor. To call that star "sun" is already a metaphor. This seems obviously just as arbitrary as to call that same object Sol, Helios, Amaterasu, etc.; in fact it is not all that arbitrary: it follows the thread of a certain history, where "sun" comes from "sol" and not "hêlios," but where "sol" and "hêlios" are both derived from the same Indo-European SWOL, and so on. When one goes back into this Cuvier-like branching (§ 41) traced by human history (in this case, the trajective history of languages), you can see for example that the flower marigold, although it has lost any link to "helios", refers to gold, linking it to the color of the sun.

– One can at least say that this flower is yellow like the sun.

– Yes, but it isn't just a question of words or colors. And it isn't just about human history; on another time scale, it also concerns the evolution of species. The human eye grasps electromagnetic waves of 578 to 592 nm as the color yellow, but the oak leaf grasps them as energy used to combine water and carbon dioxide to produce oxygen and sugars.

– They have nothing to do with each other!

– They do indeed! In both cases, it concerns the same principle: *grasping* S (light) *as* P (a color or energy). The link between S and P is established by a third term, the I interpreter (the human or the oak) in the S–I–P triad. The difference between those two interpretations of the same subject is not arbitrary; it is derived from a long series of junctions from the beginnings of life on Earth, but the principle remains the same: it is a trajective chain, a series of assumptions and hypostases of those assumptions. The same with metabolism: the cheetah takes on the antelope as prey, and its digestive system takes it as nutrition; the assimilation of those elements hypostatizes them in its life and hence its subjectity as cheetah, and so on. Necessity or chance in this case only feature in the abstract focus on the object (§ 14) that scorns the concrete historical thread. On the contrary, we saw the progressive establishment of a certain milieu and the corresponding human, animal or plant subjectity, enabling it to have the freedom to interpret S as something that goes beyond (*meta*) the repetition of the identical. It is the same principle that allowed Rimbaud to write: *A black, E white, I red, U green, O blue: vowels.*

– There are no yellow vowels?

– No, you have to combine black, green and white with two consonants, J and N, to get yellow [**jaune** in French], even though it is a primary color; however this is not the question. The question here is that those metaphors have the same principle as the assimilation of chlorophyll, or Rimbaud's metabolism itself. It is about transforming substances (S) into other substances (S′). Trajection is not alchemy, but has been the poetics of life on Earth for 3.8 billion years. From metabolism to metaphor, it is the Earth's poetics, a peculiarity that does not happen on Venus or Mars, at least not to the same degree.

This poetics does not go just anywhere but takes as we have seen a certain direction. Its path does not follow some sort of teleology, but Machado's route (§ 41): it becomes clear *al volver la vista atrás*, retrospectively. This has a relation to what Monod called *teleonomy*:

> The notion of teleonomy implies the idea of *oriented*, *coherent* and *constructive* activity. According to these criteria, proteins must be considered as the essential molecular agents of teleonomic performances of living beings.[28]

– Well, well! "Oriented, coherent and constructive?" Isn't this subjectity?

– No, no, don't worry, for Monod it is part of mechanics:

> The organism is a machine that constructs itself. … In the final analysis, all these teleonomic performances by proteins rest on their so-called "stereo-specific" qualities, that is their capacity to "recognize" other molecules (including other proteins) according to their *form* which is determined by their molecular structure.[29]

– Understood. So what is the difference between teleonomy and trajection?

– In teleonomy according to Monod, chance and necessity cancel freedom of choice. "Form" here is pure *topos, peras akinêton prôton*, "*the immediate immobile limit* [of the thing]" (§ 6) with a restricted identity. In other words, neither a *chôra* relativizing it, nor a subject that would interpret the latter and make it *ek-sist* outside its identity through choresy (§ 22). On the other hand in trajection, the relation S–I–P *necessarily* implies a degree of freedom, and hence creativity, of the interpreter (I), since P is never the strict identical reproduction of S. "I" is the one *choosing* historically and evolutionarily to metabolize S as something different from S, that is S′, and again as something different from what another interpreter would choose. The choice is not arbitrary. It is not made at random, but following some direction; without it, there would be no direction or branching in evolution, but chaos forever, the final entropy even before starting out. Indeed, before any statistical orientation can take place by teleonomy, it is only with small numbers that birth and life can occur. Life is *always* oriented in a certain direction through successive branching, and *never* haphazardly. It is done mathematically, as we have seen above (§ 41), from the beginning according to a certain experience (*al volver la vista atrás*)[30] on the one hand and in response to the milieu (*la vista alrededor*) on the other hand. I stress the *milieu* (S/P) and not just the *environment* (S); in the end it means that from the beginning, it took place according to a mediance derived from a trajective chain where the living and its milieu co-originate and co-anticipate each other.

– In short, you take up an idea already professed by Auguste Comte (§ 35)?

– Yes, in a certain sense; but adding subjectity changes a lot of things. Between Robin's mesology and mine, Watsuji and Uexküll have intervened, plus the concept of trajection, and the formulation of the trajective chain (((S/P)/P′)/P″)/P‴ ..., where S is indefinitely assumed as P, and S/P is hypostatized in S′. At each stage of the chain, the living as subject (I) has the freedom to choose case by case between various Ps that go beyond the mechanical iteration of S (i.e. necessity). The taking up of S in P raised life above matter, the biosphere above the planet and later, with technology and symbols, the ecumene above the biosphere. This is what produces evolution, rather than the indefinite repetition of the same physico-chemical mechanisms. And later on it produces human history instead of slow evolution, or even evolution itself – who knows? P, in other words the world (§§ 8, 16) pulls evolution forwards, and history even more so.

– Pulling evolution forwards is a big job!

– Yes, and that is why the metabolic process of life in general consumes energy, and why in particular our brain's metaphors consume one-fifth of it. They carry what metabolism has already projected (*ballein* and hence *bole*: throw) even further, beyond (*meta*): simple matter to create life. At every ontological level of the living, trajection is at work. Even in the neurons!

Whether Jacques Monod's spirit likes it or not, what differentiates this work from a machine's is precisely P's contingency (interpretation), in other words I's freedom (the interpreter's). This freedom, to be sure in different degrees and always relatively, exists already at the level of the gene. This is shown by the "epigenetic revolution" that overthrew neo-Darwinian orthodoxy. For example today we know that the changes in the relation of the organism to the environment (food, medical care, etc.) can modify genetic expression, and even be transmitted to the next generation.[31] This leads to comparing the genetic code to a book whose reading differs according to the moment: "epigenetics makes possible several readings of a fixed matrix (a book or the genetic code) giving rise to diverse interpretations".[32] Is this anything other than the principle of S/P trajection? Ikeda Kiyohiko for instance speaks not just of the genome, but of "an interprettive system of the genome" (*genomu no kaishakkei*) in which heredity might not be Mendelian,[33] but where the function of genes would be determined "case by case" (*sono ba kagiri*).[34] Indeed, there have been some cases of a transmission of characteristics epigenetically acquired by the modification of the interpretative system of the genome.[35]

– That is indeed contingency!

– And yes, that is life!

Notes

1 The *môlu* is supposed to have been a sort of garlic. In the 2002 French edition of the *Odyssey* (Paris, Les Belles Lettres), annotated by Sylvia MILANEZI, she specifies on p. 113 that according to Theophrastus in *The History of Plants*, IX, 15, 7, the *môlu*, used at that time as an antidote and in magic operations, grew around Pheneos and Cellene, that is, in Arcadia, the land of Pan, symbolizing nature.

2 John GRIBBIN, *Simplicité profonde. Le chaos, la complexité et l'émergence de la vie* (Profound simplicity. Chaos, complexity and the emergence of life), Paris, Flammarion, 2006, p. 47.

3 More precisely, two piles of a billion atoms in one, and a billion in the other. On this experiment, see *Sciences et avenir*, 793, March 2013, pp. 50 ff.

4 For further reading, see Luciano BOI, *The Quantum Vacuum*, Baltimore, Johns Hopkins University Press, 2011, and Michel BITBOL, *De l'intérieur du monde. Pour une philosophie et une science des relations* (On the world's interior. For a philosophy and science of relations), Paris, Flammarion, 2010.

5 However, no matter how much of an object it is, this device presupposes the existence of a human subject making it and using it for a certain objective. Science itself cannot transcend our world and whatever it finds continues to deploy the ecumene.

6 The title of chap. 5 in his *Qu'est-ce que le vivant?* (What is the living?), Paris, Seuil, 2012.

7 *Op. cit.*, p. 87.

8 *Op. cit.*, p. 86.

9 *Op. cit.*, p. 132.

10 Jacques MONOD, *Le hasard et la nécessité. Essai sur la philosophie naturelle de la biologie moderne*, Paris, Seuil, 1970. (*Chance and Necessity. An Essay on the Natural Philosophy of Modern Biology*, New York, Alfred A. Knopf, 1971.)

For a recent discussion of those theses, SATÔ Naoki, *40 nen go no "Gûzen to hitsuzen". Monod ga kaita seimei, shinka, jinrui no mirai* (Chance and necessity, Forty years later. Life, the evolution and the future of humanity according to Monod), Tokyo, Tôkyô Daigaku Shuppankai, 2012.

11 Today we might add "intelligent design", not to mention creationism.

12 I found this famous poem by MACHADO (1875–1939) on the Internet at "Caminante, no hay camino".

13 François JACOB, *La logique du vivant. Une histoire de l'hérédité* (The logic of the living. A history of heredity), Paris, Gallimard, 1970, pp. 329–330.
14 Hervé ZWIRN, Énumérer la vie (Listing life), *La Recherche*, 365, June 2003, p. 104.
15 According to *Le Robert, dictionnaire culturel en langue française*, entry "embranchement".
16 This theory was introduced by Stephen Jay Gould and Niles Eldredge in an article published in 1972 (Punctuated equilibria. An alternative to phyletic gradualism, pp. 82–115, in T.J.M. SCHOPF (ed.) *Models in Paleobiology*, San Francisco, Freeman, Cooper): it postulates that evolution includes long periods of equilibrium or quasi-equilibirum, punctuated by brief periods of important changes such as speciation or extinctions (Wikipédia, April 28, 2013). This idea may have been anticipated by Pierre Trémeaux in his *Origine et transformation de l'homme et des autres êtres* (Origin and transformation of man and other beings) (1865) and from there the later Darwin.
17 Josef REICHHOLF, *L'émancipation de la vie*, Paris, Flammarion, 2009 (*Der schöpferische Impuls* (The creative impulse), 1992), p. 199.
18 *Op. cit.*, p. 291.
19 *Op. cit.*, p. 208.
20 Michel GRIFFON, *Qu'est-ce que l'agriculture écologiquement intensive?* (What is ecologically intensive agriculture?), Versailles, Quæ, 2013, p. 46.
21 It would be better to call it Spencerism, since the most important ideologue was Herbert Spencer; even as Darwin himself fought this evil ideology, it applied Darwinian concepts to human societies (struggle for survival, survival of the fittest), and later led to racialism and eugenics or even Nazism. From a mesological perspective, this ideology is absurd since human reality is not only biological (or ecological) but no less technical and symbolic.
22 Stephen Jay GOULD and Elizabeth VRBA, Exaptation. A missing term in the science of form, *Paleobiology*, 8, 1982/1, pp. 4–15.
23 Stephen J. GOULD, *La structure de la théorie de l'évolution*, Paris, Gallimard, 2006, pp. 1718–1725 (*The Structure of Evolutionary Theory*, 2002).
24 Gould, *op. cit.* p. 1721.
25 The name of an episode in *Alice through the Looking-Glass* where Alice and the Queen must run to stay in place. On this hypothesis, proposed by Leigh Van Valen, see Pierre-Henri GOUYON, Jean-Pierre HENRY and Jacques ARNOULD, *Les avatars du gène. Théorie néodarwinienne de l'évolution* (The gene's avatars. A neo-Darwinian theory of evolution), Paris, Belin, 1997, p. 292.
26 The most famous example is the coelacanth, a group of crossopterygian fish that have changed little over the last 350 million years. Traditionally, the champions were the lingules, supposedly 550 million years old. However, according to Christian Emig, research directeur at the CNRS (see "Lingule" online)

> While the Lingulides constitute a panchronic group, the bradythelic evolution within a group that was relatively little and poorly fossilized, remains to be evaluated. However, recent anatomical studies (Emig, 1986, 1990; Biernat & Emig, 1993; Jin et al., 1993; Kowalewski & Flessa, 1996) show that recent and actual lingules have significant evolutionary differences, even as they have retained an ancestral life style. They can no longer be considered as living fossils.

They do not seem to be running as fast as Alice, to say the least.
27 Monod, *op. cit.* in § 41, p. 67.
28 *Op. cit.*, p. 67.

29 *Op. cit.*, p. 68.
30 Or, on this side of the Pyrenees: "when one looks back; and further: the view all around".
31 On this theme see Andràs PALDI, *L'hérédité sans gènes* (Heredity without genes), Paris, Le Pommier, 2009; IKEDA Kiyohiko, *Shinkaron wo kakikaeru* (Rewriting the theory of evolution), Tokyo, Shinchôsha, 2011.
32 Paldi, *op. cit.* p. 140, quoting Thomas Jenuwein.
33 Ikeda, *op. cit.* p. 119.
34 *Op. cit.*, p. 136.
35 *Op. cit.*, pp. 73–74.

10 History, evolution, trajection

§ 43. The lemmic of life and the birth of the subject

Let me borrow the title of a great book, the *La logique du vivant* to pay tribute to the author François Jacob[1] (*The Logic of Life. A History of Heredity*, translated by Betty E. Spillmann, Princeton University Press, 1993). But I want to add another perspective that was not available when it was written almost half a century ago, which concerns the trajectivity of things, both in human history and in the evolution of species. In mesology space and time are not just defined by physics, but include the concrete milieu and history reciprocally embodied in a process – life's trajection – whose concreteness by definition cannot be grasped by the abstract logos. Life is necessarily concrete or rather concrescent; otherwise it would decompose at once, leaving no organisms. It necessarily combines, for and by a certain subjectity, a certain history in a certain milieu. This is the meaning of concrescence.

Life's mediance and trajectivity cannot be reduced to *the* logic even if they do have *a* logic. That logic, irreducible to *the* logic, is a lemmic logic, in other words a "middle logic"; not so much half-way between A and non-A, but rather both at the same time. It starts with the illogical truth that there is life only through death because life creates and always engenders death. In short, through a *syllemma*, life is/is not death. Life through death, death through life: this is not mysticism, but more than biology;[2] it is mesology that takes into account the living's mediance and its being-towards-life (§ 24) that necessarily passes through death.

Even if it is not mysticism, it is at least a poetics, taking into account what creates (*poiei*) the new, going beyond the same or its repetition. It makes history possible and the evolution towards something other than the identical. This is what scares the logos and led Plato to exclude poets from the city (§ 32)[3] ... However, we need to confront this rationally. Mallarmé will always ask the question:

> Le vierge, le vivace et le bel aujourd'hui
> Va-t-il nous déchirer avec un coup d'aile ivre

Ce lac dur oublié que hante sous le givre
Le transparent glacier des vols qui n'ont pas fui!

Here is the sonnet in Barbara Johnson's translation:

The virgin, vivacious, and lovely today—
Will it rend with a blow of its dizzying wing
This hard lake, forgotten yet haunted beneath
By the transparent glacier of unreleased flights!
(https://chs.harvard.edu/CHS/article/display/6643)

But the fact is that for 3.8 billion years, life's dizzying wing beat never stopped tearing up identity, that forgotten hard lake; even while respecting it, through heredity! How can life be both itself and its contrary?

This syllemma is the same as that of the one and the many. Why does the phenomenon of life, which is one, require the multiplicity of beings and their variety? To be sure, the Buddhist formula *issoku ta, ta soku ichi* 一即多、多即一 assures us that "the one is the many, the many is one"; however, to abide by such an aphorism alone would be to reject any possibility of measurement, hence any science. Could there be a scientific version of the same principle?

At least it is clear that downstream from the origin, diversity engenders diversity. This is how a new life form emerges, of which no intermediary evolutionary form can be found: lichen were born from an association between algae and mushrooms.[4] This raises the question of symbiosis, posed by Reichholf at every scale: between cell and mitochondria (hence the animal cell), cell and chloroplast (hence the plant cell) organism and gene: "Life is both metabolism [of the organism] and the storage of information [by genes]; neither of those two elements taken separately is enough."[5] This calls to mind mediance! Those symbioses are partnerships: "the organisms are not puppets of the genes",[6] contrary to Dawkins' view (§ 37) which is again a dictatorship of identity, the very principle of mechanism. For Reichholf, on the contrary, life is oriented towards ever increasing emancipation which among other aspects involves the relation of the organism to its environments: "The forms that survived in the course of evolution have always been those that pushed furthest towards the progress of metabolism and the degree of emancipation with respect to environmental conditions."[7]

This is probably only half true. Here Reichholf goes too far, for the life of most evolved organisms always rests on that of the most primitive. It is rather the case that life at the same time maintains primitive elements while continually innovating. It does not change but it changes; life is both continuity (the one) and discontinuity (the many), as Imanishi from the beginning emphasized in Nishidian terms.[8] Again, this must seem contradictory. It is certainly illogical, but quite syllemmic. As in Machado's poem (§ 41), life can change, find new ways, by the very fact that while the path left behind can no longer change, it can be relied on in the future. This is rather commonplace, and in numerous disciplines, including the theory of evolution, it is reflected in the trivial expression "the ratchet effect"; it means that the chosen path cannot be reversed and that therefore one can go forward into the unknown.

Mesology has a more sophisticated version of this ratchet effect: trajective wedging (§ 34), and an even more elaborate one, the trajective chain, which indefinitely alternates assumption and hypostasis (§ 35). Without going back over this, I would like to draw the following inference:

In a sentence of the *Confessions* discussed above (§ 6), St. Augustine writes: "You remain in my conscience [Lord] and that is where I find you." What is usually translated as "conscience" is in the Latin text *memoria*. The word *conscientia* already existed, including in today's meaning; why then did he write *memoria*? There is actually a link between the two terms: by looking back on one's past, as Machado's traveler did, one becomes conscious of that which for St. Augustine signified God's presence; for an ordinary being (human or non-human) *memoria* signifies the meaning of things with respect to the self and therefore consciousness of one's own identity. Without the *vistas atrás* of memory, there is no consciousness of one's self and hence none of the subject. The inference is that a subject exists as soon as there is an association between a memory (of the gene for instance) and a reference external to that memory (its location.) In other words, where you have a milieu and a history, a relation between them generates a subjectity: the interpreting authority.

The problem then is to situate that authority (i.e., the subject in question), as well as the degree of subjectity or consciousness. According to this hypothesis, there may be several scales and levels. Encouraged by Imanishi (§ 35), I am no longer afraid to call a spade a spade, and so I will call "subject" what Jacob called "integron", which he defined as a certain level of integration in "the tiered architecture ... governing the construction of every living system".[9] It goes without saying that making such a connection is perfectly heretical in the Cartesian tradition which for instance still leads Alain Prochiantz to believe that animals are not subjects (§ 41). For mesology however, "integration" means both the memory of a certain self-consciousness and the capacity to distinguish oneself from the surroundings, in other words, subjectity. According to the same principle, this is where I find the ancestor of the subject rather than Jacob's integron: "The ancestor could at first be just a kind of core, the association of a few molecules helping each other to somehow regroup."[10] Immediately thereafter, Jacob asks the following question:

> But then, how did it begin? The genetic message can only be translated by the very products of its own translation. Without nucleic acids,

proteins have no future. Without proteins, nucleic acids remain inert. Which is the chicken and which is the egg?[11]

I see this as a biological version of the self-predication that expresses and activates especially human subjectity. (§ 8). Jacob does not answer the question he poses: "Which of the polymers, the nucleic or protein, comes first? What is the origin of the genetic code?" Similarly, mesology cannot say what comes first, the origin of substance or that of interpretation, of S or P; to answer it would be a mystical leap, because to ask the question is (always already) only possible downstream from the S/P trajection (§ 34).

– And so, do you see proto-cells of the subject, culminating in the *cogito?*

– Yes, and even beyond, but of course with a scale in degrees of subjectity. As with electromagnetic waves that are only visible in a certain range but nevertheless exist on either side, there is subjectity below and beyond our consciousness. Below, as we just saw. Beyond, and according to the same principle, i.e. at a certain level of integration, we postulate, like Imanishi, that there is subjectity up to the level of what he calls *seibutsu zentai shakai*, "the whole biotic society" (§ 35). This seems less artificial than the "biotic citizenship" mentioned by J. Baird Callicott,[12] or *a fortiori* the "assembly of non-humans" discussed by Bruno Latour.[13] The issue is not to pin human institutions on the living world, leaving aside any metaphor, but on the contrary to take into account the essence of the living, of which we are but a particular instance. Unlike matter, the essence of the living is to create its own world and to be its subject.[14] It is the subject whose subjectity necessarily impregnates its milieu by trajective sequencing, at every level including that of the geo-cosmos, by virtue of the double principle of the Earth's poetics (§ 36). I believe this is the sense in which we can speak of the subjectity of a species, like Imanishi, and propose a *bio-hermeneutics*.

Here we find Uexküll again, and Watsuji as far as the human is concerned, but their problematic has expanded and become more complex. Without going into the details of such a bio-hermeneutics, which emerges from animal and plant ethology, we can see four principal levels of integration or subjectity: the cell, the organism, the species and life as a whole. Life as a whole creates a world where species can create their own worlds, and similarly the species for the organisms and the organism for the cells. The other way around however, the organism can only create its world with cells etc. This is not causality, but trajection, because at every level and in both directions interpretation occurs, namely the creation of meaning. We remain in the ternary schema S–I–P (subject–interpreter–predicate § 40), indefinitely pursued in trajective chains (((S/P)/P′)/P″)/P‴ ... etc, always according to the model of the syllemma where S is/is not P (i.e. S as P).

– In short, nature writes messages whose writing the living must decipher?

– Yes, besides, it is a very old idea, going back well before Galileo and his famous "Il libro della natura è scritto in lingua matematica" ("The Book of Nature is written in the language of mathematics"). In Alain de Lille's book The *Natura* mentioned before (§§ 16 and 24), nature gave Venus the task of writing the world.

> To carry out her writing task, I [Natura] assigned her [Venus] a very powerful pen: on the appropriate pages that needed to be written and which I had generously put at her disposal, she had to represent the genre of things and comply with the orthography; I feared that she would not prevent her pen from straying into the sideways of misspelling outside the path of correct description.[15]

– Of course, those "transcriptases"[16] must do their job correctly! If not, watch out for mutations!

– For sure! The DNA is in the position of S, the RNA of P and the enzyme of I. However, in the case of subjectity, we have something better than enzymes. We do not want to brag, but it is in humanity, in other words in the ecumene and history, that trajection has gone furthest. This is related to the fact that subjectity culminates in the human, both objectively (as cognitive science has shown), and subjectively, since we see the other worlds from our own standpoint.

§ 44. The world's Cave and the desert of the logos

As we have seen (§ 35), the S–I–P ternary has similarities with Peircian semiosis; however it is the Japanese language (at least among those languages I have surveyed) that embodies it most clearly. While the main European languages only have the abstract S–P dyad ("Hanako is sad", or "S is P"), Japanese must imply the concrete existence of the speaker ("Hanako seems sad", or "S is P for I", § 4). In linguistic terms, this non-*ex*plicit *im*plication of the speaking subject in the scene of the utterance is associated with the dissemination of his subjectity in the scene, which makes him into an *ambient* rather than a subject in the sense with which we westerners are familiar (§ 6). In so doing the opposition subject–object that founded the classic modern western paradigm is lessened (§ 9); other cultural legacies from Japan have favored this tendency, in particular Buddhism with the idea of transmigration and the tetralemma (§ 9). In the 20th century this synergy produced Imanishi's "naturing" science (*shizengaku*) (§ 38) in which the subject–object, human–nonhuman opposition tends to vanish as subjectity is extended to the whole living world.

This naturing science is not only alien to dualism, but also to nominalism which, with the arrival of the MOT, has dominated western modernity, especially in Anglo-Saxon countries. A culture in which it is possible to declare that "there is no such thing as society" (§ 12) is *a fortiori* not ready to admit that something like speciety (§ 38) could exist. That culture tends to favor methodological individualism and prefers to follow those who speak of the selfish gene (§ 38), which shows that it reflects the MOT just as faithfully as Imanishi's argument reflects the Japanese ambient. On both sides, all this is part of the worldliness (*Weltlichkeit*) of human affairs, in other words, of Plato's Cave. When Dawkins' mechanistic nominalism leads him to write that altruism is determined by the gene's selfishness, he is just as worldly (*weltlich*) as is Imanishi in his vitalist realism[17] which he expressed as follows:

> In the long run, the species is transformed by itself (*mizukara* 自ら) spontaneously (*onozukara* 自ずから)into a different one. As a result, the origin of species is in the species, and not in individuals.[18]

Or:

> [Similarly], if the baby stood up, it is because he had to stand up (*tatsu beku shite tatta*)' … evolution has evolved because it had to (*kawaru beku shite kawatta*). … To say that it changes because it has to change is to see evolution no longer from a mechanistic point of view, but as a course (*Kôsu*).[19]

As for knowing what really happened (S), all we can say at this point in human history, is that it is the MOT (à la Dawkins) that is well established, i.e. dominates P predicates, and not the ambient (à la Imanishi), including in Japan where Imanishi is ostracized by the academic world. According to the principle of trajectivism, consistent with the fact of mediance, mesology leans more to Imanishi's views than to Dawkins'. That said, I will not go as far as to liken Imanishi's course to Machado's "path" mentioned above, because the former does not seem to be limited to a retrospective view. His view has a whiff of finality reminiscent of Imanishi's own subjectivity; one could make the same criticism of astrophysics, with the idea of the "anthropic principle" or of "intelligent design". It seems to me that the trajectivism of the Japanese ambient leads him to the same error as Watsuji when his mesology lapses into determinism: to confuse hermeneutics (trying to understand another subjectity from the inside) with introspection (deducing another's subjectity from one's own subjective impressions).[20] I see here the weak point of a trajectivism that eliminates the protocols of positive science, namely the impossibility of ever transcending one's own worldliness. That is the very reason why modern science was not born in Asia but in Europe.

Nevertheless, the fact remains that whether it concerns evolution or history, it is impossible to grasp S (past events) in any other way than with a worldly predicate (P) that belongs to the present, in other words, inevitably in the twilight of the Cave. As its most well-informed representatives know,[21] biology has become a historical science when dealing with evolution. This is the very criticism that Robin addressed to Darwin.[22] The fact that today biology can produce phosphorescent sheep does not change the reality that experimental proof about the past is unobtainable. This makes it necessary to understand the logic that supports our predicates in such improbable and non-experimental matters.

This is the same logic of worldliness, described by Nishida as the "logic of the predicate" (here lgP) which he opposed to the identity of the subject (here lgS). The latter can be illustrated by the traditional syllogism: "All men are mortals / Socrates is a man / Therefore Socrates is mortal." This syllogism is logical in as far as the subject of the minor term (Socrates) is included in the major term (mortals). Here we have the identity of the subject.

– Major, minor, all this is tough! Couldn't you wrap it up a little?

– Yes, of course! Just listen to Nature's complaint:

> The major, which is on top, has the task of predication, while the minor is subject to the laws of domination. In the first proposition, the predicate is inherent in the subject, not in the way of a real inherence, but because of an exterior link, just as when one term predicates another. As for the second proposition, the major on top is intimately linked to the minor thanks to the kisses they exchange. But in the conclusion, where the inherence is perfectly expressed, the connection operates thanks to a true link that unites the subject and the predicate in their flesh.[23]

– I understand! The union of the subject and the predicate in the flesh, that is the S/P reality!

– Certainly! As for the identity of the predicate, I will illustrate it with the following example which will show you that it is indeed the essence of worldliness: "A successful person wears a Rolex / I am wearing a Rolex / Therefore I am successful." Here the subject of the minor is not included in the major; it is only juxtaposed in a going-together-with (concrescence) embodied by the identity of the predicate "wearing a Rolex".

From the point of view of the logos (lgS), the second example is of course absurd, but it is a very human reality, and to reject's one worldliness, lgP in other words, would be a form of self-righteousness, which is the fount of all perils, as Pascal would have said. The most prudent thing is to recognize that human reality is really part of worldliness, at least to a certain extent.

While the worldliness of our native languages, for example the English language, leads Dawkins to make objectiform utterances (if not more objective that Imanishi's) of the binary type "S is P", the Japanese language on the other hand leads Imanishi to subjectiform (if not more subjective than Dawkins) utterances of the ternary type "S is P for I", where the I is a diffuse and preferably collective subject at every level of life, as befits the Japanese ambient. Dawkins on the other hand individualizes his object at the most elementary level, as befits the British MOT. We are back in the twilight of the Cave, with both!

– Wait a minute! You forget that science, born of European modernity, comes precisely from the Enlightenment that western philosophy sought outside the Cave!

– To leave the sensible world (the Cave) does not suppress the sensible world; it means choosing abstraction – the liberation of the logos from existence (§ 31). It is as much a foreclosure as a liberation. It means going to the desert, in a de-serting (an unraveling §7) of the links that make a world, foremost the world of life, the *Lebenswelt*, and even the geo-cosmos (§ 36). When Descartes, in his *Principles of Philosophy*, abstracts "pure science" from "sentiment", in other words from the sensible world, he accomplishes symbolically a move similar to that of the Fathers of the desert like St. Anthony, who through anachoresis rejected their medial body, just as they rejected their animal body through asceticism. By leaving the city for the desert, the ecumene for the wilderness, "drunk on God", according to Jacques Lacarrière,[24] they intended to detach themselves from the mediance of human existence itself. In both cases – pure science or asceticism – it means choosing the search for the absolute, at the expense of the relative. However, concretely, life, material things, our nature and our bodies are relative. Very logically then, Climaque could write: "The separation from the world is the willful hatred of matter and the negation of nature by desire to reach the supernatural."[25]

St. Anthony's or Climaque's desert prefigured at the individual scale, impelled by faith, what on the worldwide scale, impelled by technology, finally produced modernity: the refusal of our mediance which binds us to others, to life and to matter, in other words, to the Earth. In either case, it means trying to be angels while we remain animals. Or rather – and this is the third lemma – we are neither angels nor animals but we are also both at the same time, and so here we have the fourth lemma.[26]

– But how can we negate our mediance?

– Obviously such negation is only virtual, since the structural moment of human existence can only be abolished by death (and only for the "I myself"!). This does not invalidate the principle of a negation which substitutes an absolute for mediance. Such an intention did not originate with Christianity: it was already present in Platonic metaphysics symbolized by the myth of the Cave. In the onto-cosmological terms of the *Timaeus*, it would amount to abstracting the relative being (the *genesis*) from its *chôra* (the milieu, the sensible world) in order to merge it with absolute being (the *eidos*).[27]

However, one cannot live in the light of the desert; concretely, without the Cave sheltering our existence (the scientist's own life first of all), there could be neither meaning, nor the light of science. The Cave is shadows, the desert is light, but reality is the chiaroscuro of the syllemma which *comprises* both. Trajectively: the one as the other, S as P.

– So, one should not bother to choose?

– It is not that simple. To choose (*legein*) is what the logos does, and not only the logos. This is also what life does when from the surplus of information contained in the biosphere, it selects a certain amount to metabolize into the meaning of a certain species (§ 36). Similarly, among the surplus of information in the environment, human life selects a certain amount, to anthropize- humanize it as the meaning of a certain human milieu. Without such selection, there would be neither life nor culture nor logos, and thus it is necessary.

– Aren't we returning to Husserl's argument (§ 29)? Some science, yes, but no science that would not presuppose the *Ur-Arche*, that fundamental origin of ours, the Earth.

– In a certain sense, you could say that. In spite of all its protocols, science will never transcend our existence. It can transcend *a* world, (what we call a paradigm shift, which amounts to a change of predicate) but it cannot transcend worldliness as such. That is why nature seems to be written in the language of our mathematics. To be sure, for science in the end, the highest abstraction unites the laws of the Universe and our laws. This is what Plato wanted to say and what Roger Penrose also says (§ 34). That is the ideal of the logos; however it is not how Husserl understood it. On the contrary: in the end, even science cannot eliminate our belonging to the Earth, in other words to the ecumene.

– And what do you think?

– This is also what I think. In mesology, to go beyond trajective human reality (S/P) and so reach the absolute (S), requires a mystic leap (§ 34). To absolutize the S subject, as science does, is no less mystical and less absurd than to absolutize the P predicate, as religions do or as the Kyoto School has done (§§ 8 and 27). Or more concretely put, what scientism and fanaticism do. One could even say that today they feed each other. For mesology on the other hand, in either case, there can only be an orientation – a *tension-towards*, and no more, the *qu* 趣 Zong Bing discussed (§ 15). I would equate it here with the "as" of the "S as P". For this to be an orientation, however, it necessarily presupposes that-which-from-where we can orient ourselves to somewhere else.

Zong Bing called this that-which-from-where, *zhi* 質, "substance", which the aesthetic tradition wants to translate by "material forms", but which I equate with S. However, we cannot live in the spiritual abstraction (*ling* 靈), towards which, according to him, the landscape tends, which I would equate here to P. But if this were the only dominant truth, just as if the in-itself- only of S were to dominate, there would no longer be a tension-towards. In other words, nothing would have any meaning, and there would be no life, and therefore no more possible science. We owe the possibility of tension-towards to the structural moment of our existence: our mediance as humans, and especially as living beings, which attaches us to a milieu, a history and finally to the Earth (S) in the relation of the Earth (S) to the sky (P), which is the principle of all cosmicity (§ 19).

– I see, but I don't see how to choose to tend towards S or P … Could you, in concluding, summarize what mesology can tell us about this?

§ 45. *Natura natura semper*: poetics of the obscure Female

– The choice of tending toward this or that is attributed primarily to human beings. It is specifically what is called since St. Augustine free will. Such a perspective is opposed to determinism and fatalism. In the West it has become ingrained in the notion of the subject, so much so that it became consubstantial with the MOT. This enabled science to free itself from religion by choosing S over P. This in turn did not prevent science from also falling into determinism, in other words the contrary of free will, as for example in the theory of the selfish gene (§ 37). Mechanicism is fundamentally deterministic. Nor did it block chimerical combinations between the post-religious label for free will, arbitrariness and mechanicism, as in the structuralist theory of the origin of meaning and language (§ 27).

These incoherent theories result from dualism which, by foreclosing mediance, cannot see how this structurally condemns it to recurring conflicts between the subjective and the objective. The MOT subject and the MOT object are in fact co-originated at a given historical moment (modernity) and by the structural moment of human existence (§ 13). This means that the modern subject and the modern object are co-anticipating (*sôdaiteki* 相待的, § 34): they anticipate each other to be what they are.

For mesology on the other hand, the capacity to choose is inherent in life from the start, when life takes a certain direction as a result of certain choices. I insist, *in a certain direction*, and not towards a certain goal, whether it be called the "Omega point" according to Teilhard de Chardin,[28] or something else. Those "goals" are due to a mystical leap. Mesology on the contrary believes that the direction is the one taken historically by the *senda que nunca / se ha de volver a pisar*, the path one never has to tread again (§ 41). It is traced by us living beings, and orients us; it is the one we follow every instant towards the unknown.

– Tell me, is this not determinism or fatalism?

– No, because at every moment, the path can bend one way or the other, according to our choices and circumstances. Let me emphasize *circumstances*, and not chance. Indeed, the path already traveled determines contingently that everything that comes up already has a direction. Very simply, because given the subjectity of living beings, and *a fortiori* of humans, we exist in a certain milieu following a certain history. Circumstances appear concretely in such a milieu and such a history that necessarily presuppose our existence, as well as its capacity to choose that our subjectity no less necessarily includes. They originate in us and the choices we constantly make in the trajective chain that continues indefinitely since the origins: (((S/P)/P′)/P″)/P‴ … and so on.

– Are you saying that when I choose to do this or that, I just repeat what life was already doing more than 3 billion years ago?

– No, not "repeating": creating anew, every time and according to your own nature; but always following the orientation that life gave itself from the beginning. From the beginning, it chose to *ek-sist* beyond the identity of matter, which never chooses precisely because it has no subjectity, nor a world in which to *ek-sist*. It is *weltlos*, without world and therefore incapable of choosing. Nevertheless, life appeared and developed on Earth, and perhaps elsewhere (but where, we do not know at this point), and with it the capacity to understand things (S) as something (P). This resulted from a very ancient *auto-poiêsis* following a direction in space and time[29] (§ 35) embodied in the deployment of the Universe and related to what William Golding, for James Lovelock, has named "Gaia".[30] In other words, it was the beginning of the trajective chain, whose every link in turn creates a new way of understanding things: P′, P′, P‴ and so on, while transforming itself: S′, S″, S‴ … and so on, in the course of evolution and history.

– Isn't it contradictory to speak of "the beginning of the trajective chain"?

– Of course it is contradictory. We can never get beyond a trajective wedging, where S is always already the hypostasis of an S/P. Thirty years ago I finished *Le sauvage et l'artifice* (*Japan. Nature, artifice and Japanese culture*) with a chapter entitled "Nature, the ultimate subject",[31] but today I have to see nature as the hypostasis of an S/P where our subjectity has the role of I. That is precisely the error of any hypostasis: to confuse a subject with a predicate. It is indeed an error from the point of view of the logos, but it is also the principle of all reality. It is what Natura, in Alain de Lille's book, calls "the real link which ünites in their flesh the subject and the predicate" (§ 44). So if Natura tells us this, then …

– Isn't it what you called earlier (§ 18), the "worlding of the body and the somatizing of the world?"

– Yes, it is similar to what Imanishi called the "subjectivation of the environment and the environmentalizing of the subject" (§ 37). It takes up the logic already glimpsed by Auguste Comte (§ 35), but with an essential difference: the assumptions I am talking about, which as predicates create the predicate-world of a certain species or a certain culture, are sense data presupposing a subjectity and not just mechanical effects. The subjectity which is inherent in the S/P reality, and necessarily rests on the ternary structure S–I–P, not only permeates the judgments that I-here-now make of the past (history or evolution); it is also inherent in the very unfolding of history and evolution as trajective chains (((S/P)/P′)/P″)/P‴ … This means that S here always has the ambivalence inherent in the word subject itself (§ 5). S is not only both the subject of the logician and the object of the physicist, *but also* (the fourth lemma) the interpreter I, in other words it is the subject that creates S as P (§ 40) from the very origins of life. That is possible because the trajective chain is not just focused on the object (§ 14) – *focused on* S – but on the process of a real *genesis*: S/P indefinitely continued as a chain, a bit like the chain poems (*renga* 連歌) in the Japanese tradition (§ 17).

– So for you, evolution and history are a sort of *renga*?

– Yes, and that is why I speak of the "poetics of the Earth". It is not, as Macbeth said, "A tale / Told by an idiot, full of sound and fury, / Signifying nothing", but rather a chain poem. For me this is more than an image; it is a heuristic principle and a reality, the very principle of reality as S/P. Therefore I think we need to look again at what Reichholf said about metabolism in the light of Uexküll's *Bedeutungslehre* (§ 35). Evolution not only concerns the relation between organisms and their environment, but also between milieu and subjectity. It is not a question of mechanics, but of co-awaiting and co-originating.

– Do you want us to rely on Great Vehicle Buddhism?

– Only partly: I would like the third lemma to show us that in this matter, *neither* mechanicism *nor* vitalism (i.e. mysticism) are adequate. What we need is *both* (fourth lemma) to acknowledge the role of I's subjectity at every stage of the trajective chain, and to understand the latter as objectively as possible, namely with the methods of positive science. Concretely, this means we cannot do without a (bio)hermeneutics and a (bio) semantics based on a generalized ethology. There is more. In the trajective chain, at the ontological level of the biosphere and evolution, the assumptions of the *Umgebung* as *Umwelt* are concretely hypostatized, that is, they are embodied in the flesh of living beings. For mesology, this is the motor of evolution, and what epigenetics is discovering with the methods of biology.

– Indeed? But epigenetics does not mention subjectity!

– True, just as M. Jourdain did not think he was speaking in prose (note of translator: a character in Moliere's play, *Le Bourgeois Gentilhomme*). You cannot just throw away three or four centuries of mechanicism and twenty centuries of Christianity!

– And what is happening in Japan today?

– Are you asking whether Imanishi's thought had any influence? Yes and no. On the one hand, most of the academic world continues to ban him. Here for instance is a recent book, entitled *Why is Evolution a Philosophical Question?*[32] in which a team of nine philosophers of science, in almost three hundred pages, spectacularly succeeds in not mentioning his name once. It is as if in Germany a book on ontology were never to mention Heidegger.

One does not fool around with ostracism (*mura hachibu*)[33] in Japan! On the other hand, generally speaking there are still streams of thought that are fundamentally resistant to dualism and mechanicism in the lineage of what Niwa Fumio, speaking of Imanishi (§ 37), called "the method of the Japanese vision of nature".[34] This is also illustrated for example in the bio-history (*seimeishi* 生命史) mentioned by the biologist Nakamura Keiko. I believe that her perspective could be compared to Imanishi's "naturing science" (*shizengaku*, § 38) although she actually does not lay claim to this particular filiation.[35] Finally there are those who still identify with Imanishi's cause in the duel opposing him all his life to Darwinism; but they develop his theses with a few amendments. This is for instance the case with Mizuhata Shôzô.[36] The question revolves around the concepts of ecospecies and speciety (§ 38) that led Imanishi to one of his most heterodox theses. While, for Darwin, speciation is due to the long-term advantage of individual organisms presenting certain differences, Imanishi ended up postulating a global and catastrophic (in René Thom's sense) change in the species as speciety, a bit like fashion in human societies. However, he left open the question of the processes of such change. Mizuhata wants to resolve that issue with the following hypothesis: in a given speciety, males showing certain differences would be more attractive to females; this would eventually bring about speciation.

– Isn't that what Darwin already called sexual selection?

– Yes, but not from the same perspective, because here the subjectity of the species intervenes. Besides, Mizuhata moderates Imanishi's position in this regard. He excludes speaking of subjectity, and thus of ecospecies and speciety in the case of protozoas or plants, because he only sees this as possible when there is a nervous system and the capacity of movement.

At that level, one could compare Imanishi's view to the anti-utilitarianism of Jacques Dewitte.[37] For him it is obvious that unless the conditions for survival and reproduction had been at stake, life would never have deployed such an abundance of forms. This philosopher proposes the idea that for the living as for humans, appearance is necessary for being: "social forms as well as natural forms have a dimension of gratuitousness and contingency that goes beyond any principle of necessity or utility."[38]

In this context, Dewitte mentions as an afterthought[39] a famous Taoist aphorism, "the utility of uselessness". For mesology on the other hand, this is a central question. It symbolizes the difficulty in interpreting what makes sense in a world which is not ours. It explains the penchant, in our capitalist era, for mechanicism and its derived utilitarianism to eliminate this question by reducing it to quantifiable utility, such as "time is money", although time can be an infinity of other things. To interpret natural forms requires on the contrary understanding what they might mean in the world belonging to the species in question. This is a question of bio-hermeneutics, in other words of respect for the subjectity of a species that is not ours, which by definition is unthinkable for mechanicism.

Going a step further, I will give here the ultimate reason that brought me to translate *shushakai* 種社会 by the neologism "speciety". Literally *shushakai* means "society (*shakai*) of a species (*shu*)".[40] What Imanishi means by this has no relation to what the standard theory of evolution means by "population". It is not about statistics, but about hermeneutics: to understand what links the members of a certain species together, on their own terms. How do they appear to each other, do they exist for each other? The etymology of our word "species" is important here. It is derived from the Latin *species*, whose initial meaning is "view, look" from the Indo-European root *SPEK* expressing the idea of seeing, looking, as in "spectacle" or "in*spect*", and hence the meaning of aspect, appearance, looking good, nice appearance, etc. One can see that "species" and "aspect"[41] are linked. In this connection, speciety somehow becomes the appearance of being, here the being itself of a certain species. It is what makes the members of that species exist as seen by others, and what therefore may determine their behaviors.

In the end, speciety is none other than the very object of ethology extended to bio-hermeneutics and bio-history. However, there is more: in the sense I have just given it, the term speciety must be brought together with the previously mentioned concepts of cosmophany and revealed-ness (*Offenbarkeit*) (§ 36). For mesology the question is inseparable from that of worldliness, in the ontological sense Heidegger gives to *Weltlichkeit*.

I would like to retain the ambivalence of the term worldliness,[42] in both the philosophical and in the everyday meaning of frivolity. In every human society, the worldliness of high society, the leisure class[43] is what orients the behaviors of the classes right below it and so on, downward. This is what moves culture forwards. In other words the P predicates that open up a world, are those of the elites[44] who invent them, and then others imitate them and so on, according to speciety.

The question is thus linked to the very principle of the institution of all reality, the trajective chain (((S/P)/P')/P'')/P''' ... and so on. Worldliness, with all its appearances, is what pulls being forwards, for a good reason; namely, because of our mediance, we exist for one another and through each other, in each other's eyes (hence the notion of human respect[45]). At the same time, it rejoins our speciety: it insures that any member of our species, for good or ill, exists for us more intensely and more decisively than any member of another species. This is precisely how such a theory as mechanicism could emerge!

For instance, to be seen naked by a dog, not to mention by a fly, does not mean the same thing as to be seen in that condition by another person. It is according to that same speciety that the great ladies of earlier times were not embarrassed to appear in their birthday suits in front of their male servants; these menials were not a part of their world and hence beneath contempt.[46]

There is no reason except by a leap of faith, to assume that the same principle does not apply for other species. The Japanese primatological school was the first to show that speciety in that sense functions among superior primates: they imitate each other, not in everything but in what is important in their world. This opened up the door for bio-hermeneutics to extend to the entire living world. However, we are only at the courtyard of the building to be constructed by 21st century science. Formidable barriers remain as long as it is possible say that "animals are not subjects".

– In other words, mesology is impossible except by returning to Auguste Comte?

– That is unlikely because the trajective chain of history continues the evolutionary one and evolution is a fact; that is, life recreates itself ceaselessly as it perpetuates itself. As in everything, this is true of paradigms in the history of science, which are but one trajective wedging after the other. Through the lens of a "naturing science", I actually see in the magnificent image of the Latin language life's continuous self-creation as it engenders beings anew. One day, Latin speakers chose to translate the Greek *phusis* by the feminine future participle of the verb *nasci* (to be born): *natura*, "that which is about to be born".

– Why the feminine?

– Probably because *phusis* is feminine, but mostly because to give life through birth (*genesis* as Plato said of beings in the sensible world) is a feminine power. The same thing is said about the engendering power of nature on the other side of the world, in China, even though there is no feminine or masculine in Chinese:

谷神不死	*Gu shen bu si*	The genie of the valley does not die
是謂玄牝	*Shi wei Xuanpin*	It is called the obscure Female
玄牝之門	*Xuanpin zhi men*	The door of the obscure Female
是謂天地根	*Shi wei tian di gen*	Is called the root of heaven and earth
綿綿若存	*Mianmian ruo cun*	Like a spinning thread it endures
用之不勤	*Yong zhi bu jin*	Using it does not exhaust her.[47]

The door of this "obscure Female", Xuanpin, is none other than what Courbet represented in his *Origin of the World*. It is much more than a simple representation, much more than just the human history of our relation to nature: all of natural history must be reborn by itself in each of us and through each of us.

This is how nature, forever renewed at each link of the trajective chain, *ziran* 自然, "thus of itself", will always be about to be born, *natura natura semper* ...[48] particularly, to be born in all our works and choices. This is why mesology does not have a definite answer to the question you asked me earlier: to chose to lean towards S or to P will always have a new meaning, depending on milieu and on history, and that meaning will not become clear to us unless *al volver la vista atrás*, as Machado said (§ 41).

This can produce antithetical "truths" just as it can coexist peacefully in the same person.

– In short, your mesology is really casuistry?

– Yes and no (fourth lemma). Like history and with history, this science of the milieu is a *science of the concrete*,[49] and as such refuses the abstract alternative between the light of the desert and the shadow of the Cave. Both are necessary to human life, and together make up reality. The question then becomes how to marry them to make them grow together, trajectively. In my understanding, mesology is born at the crossing of modern western objectivism and Japanese-style trajectivism. It wants neither the Cave nor the desert (third lemma) but a syllemma: to take from the desert what light is needed to better illuminate the Cave of life.

Notes

1 *Op. cit.* in § 35.

2 Yes, but in acknowledging that contemporary biology, with notions such as apoptosis (programmed cell death) has indeed taken into consideration the phenomena it includes. On this theme see Jean-Claude AMEISEN, *La sculpture du vivant. Le suicide cellulaire ou la mort créatrice* (The sculpture of the living. Cellular suicide or creative death), Paris, Seuil, 1999. Even as mesology opens the door to metaphysics and ethics, it remains rooted in biology; however, just like the biosphere, it emerges from biology but cannot be reduced to it.

3 On this question and on his relation to Mallarmé, see René PASSERON, *Pour une philosophie de la création* (For a philosophy of creation), Paris, Klincksieck, 1989, pp. 107 ff.

4 Reichholf, *op. cit.* in § 41, p. 280.

5 *Op. cit.*, p. 292.

6 *Op. cit.*, p. 296.

7 *Op. cit.*, p. 297.

8 *Seibutsu no sekai*, *op. cit.* in § 37, p. 125. The initial expression is *furenzoku no renzoku* 不連続の連続, the continuity of discontinuity.

9 Jacob, *op. cit.*, p. 323.

10 *Op. cit.*, p. 326.

11 *Ibid.* The question has evolved since then, cf. Patrick FORTERRE et al., *De l'inerte au vivant. Une enquête scientifique et philosophique* (From the inert to the living. A scientific and philosophical investigation), Paris, La Ville Brûle, 2013, pp. 87–88 and 89:

> the RNA can function both as a gene and as an enzyme: they are the famous ribozymes, enzymes with ribonucleic acid. … This 1980 discovery [by Tom Cech and Sidney Altman], crowned by the Nobel prize in chemistry [in 1989], is the basis of one of the most important advances in the domain of the origins of life. It solved the chicken and egg problem: what appeared first, the DNA or proteins? The problem seemed unsolvable since proteins are needed to make DNA and DNA to make proteins. We know now that it is neither, as ribonucleic acid can play at the same time the role of the gene (like DNA) and of the catalyzer (like proteins). Obviously, the question then is to figure out what preceded organisms with RNA.

12 See in particular *In Defense of the Land Ethic. Essays in Environmental Philosophy*, Albany, State University of New York Press, 1989, *passim*.
13 See *Politiques de la nature. Comment faire entrer les sciences en démocratie*, Paris, La Découverte, 1999, *passim*. (*Politics of Nature. How to Bring the Sciences into Democracy*, Harvard University Press, 2004, translated by Catherine Porter.)
14 At another physical and ontological level, we should consider that the same principle is also valid for matter, which with the Big Bang and the Universe has created the world whose subject it is. So why not speculate that there is nothing to stop the continuous development of the same operation at other levels as well? Astrophysicists do not hesitate to do so in another vocabulary.
15 Alain de Lille, *op. cit.* in § 16, pp. 102 and 103.
16 Enzymes catalyzing the synthesis of ribonucleic acid (RNA) by transcribing the message inscribed in deoxyribonucleic acid (DNA).
17 In the opposite sense of nominalist; i.e., believing in the reality of the mind's constructions ("universals") beyond singular sensible beings. For example, society beyond individuals, or the species beyond organisms. In that sense one could say that Durkheim was a realist and Spencer a nominalist. On the tendency of Japanese culture to vitalism, see the *magnum opus* – almost a thousand pages! – of SUZUKI Sadami, *Seimeikan no tankyû* (In search of a vision of life) Tokyo, Sakuhinsha, 2007; or more briefly, his contribution to the Cerisy colloquium "Being towards Life" (2008): Formation and development of "being towards life", in WATSUJI Tetsurô, *Ebisu*, 48–49, pp. 31–44.
18 Imanishi, *Shutaisei no shinkaron*, *op. cit.* in § 21, p. 175.
19 Imanishi, *op. cit.*, pp. 202 and 204.
20 On this point, see my comments on the translation of *Fûdo*, *op. cit.* in the Preface of this book.
21 For example, Le Guyader, *op. cit.* in § 37.
22 See above, Propos.
23 *La plainte de Natura* (Nature's complaint), *op. cit.* in § 16, p. 107. On that same page, note 1 explains: "Inherence is a mode of belonging between an accident and a substance and when one follows Aristotle (*Categories*, V), there is a substantial inherence (man is mortal) and an accidental inherence (Socrates is a man)." However, Alain does not give Natura an explanation matching the theory's rigor. "Real inherence" is expressed at the conclusion of the syllogism because it is produced by the fertilization of the female subject by the male predicate.
24 Jacques LACARRIÈRE, *Les Hommes ivres de Dieu* (Men drunk on God), Paris, Fayard, 1975. On the relation between ecumene and the ereme (desert or wilderness), see my *Histoire de l'habitat idéal* (History of the ideal habitat), *op. cit.* in § 8. Eremitism is choosing the ereme.
25 This judgment by Climaque is cited by Pierre MIQUEL, in *Lexique du désert. Étude de quelques mots-clés du vocabulaire monastique grec ancien* (Lexicon of the desert. Study of a few key words in the ancient Greek monastic vocabulary), Bégrolles-en-Mauges, Abbaye de Bellefontaine, *Spiritualité orientale*, 44, 1986, p. 71.
26 Cf. Blaise PASCAL, *Pensées*, Paris, Livre de poche, 1962 (1657):

> It is dangerous to make man see too clearly his equality with the brutes without showing him his greatness. It is also dangerous to make him see his greatness too clearly, apart from his vileness. It is still more dangerous to leave him in ignorance of both. But it is very advantageous to show him both. Man must not think that he is on a level either with the brutes or

with the angels, nor must he be ignorant of both sides of his nature; but he must know both. Man is neither angel nor brute, and the unfortunate thing is that he who would act the angel acts the brute. If he exalt himself, I humble him; if he humble himself, I exalt him; and I always contradict him, till he understands that he is an incomprehensible monster. Let man now know his own value. Let him love himself, for there is in him a nature capable of good; but let him not for this reason love the vileness which is in him.

On the use of the tetralemma in Pascal …

27 The *chôra* is the condition without which the relative being (*genesis*) could not be distinguished from the absolute being (*idea* or *eidos*) of which it is only a copy, according to Plato. On the quasi-ontogenetic role of the *chôra*, cf. Alain BOUTOT, *Heidegger et Platon* (Heidegger and Plato), *op. cit.*, § 32; for the interpretation of the *chôra* as sensible world or spatial milieu, cf. Luc BRISSON, *Le même et l'autre dans la structure ontologique du* Timée *de Platon* (The same and the other in the ontological structure of Plato's *Timaeus*), Sankt Augustin, Akademia Verlag, 1994. As we have seen, I lean towards translating *chôra* by "milieu".

28 In *Le phénomène humain* (The human phenomenon), Paris, Seuil, 1955.

29 What is usually called "time's arrow" also indicates direction in space, i.e., an arrow in space-time.

30 James LOVELOCK, *Les âges de Gaïa*, Paris, Odile Jacob, 2007 (*The Ages of Gaia*, Oxford University Press, 1988).

31 *Op. cit.*, p. 285. Taken up in Marie-Claire ROBIC, Jean-Louis TISSIER and Philippe PINCHEMEL, *Deux siècles de géographie française. Une anthologie* (Two centuries of French geography. An anthology), Paris, CTHS, 2011, pp. 406–409.

32 MATSUMOTO Shunkichi (ed.) *Shinkaron wa naze tetsugaku no mondai ni naru no ka*, Tokyo, Keisô shobô, 2010.

33 村八分, which literally means eight-tenths (*hachibu*) of the village community (*mura*). In the countryside before Meiji, the houses whose members had seriously broken community rules were excluded from any communal mutual assistance except in two cases, death or fire (hence the "eight-tenths"). This included banning the use of the commons, which in practice made material life impossible and condemned the people in question to exile.

34 *Nihonteki shizenkan no hôhô* 日本的自然観の方法.

35 TSURUMI Kazuko and NAKAMURA Keiko (dialogue), *Yonjû oku nen no watakushi no inochi* (My life of four billion years), Tokyo, Fujiwara shobô, 2002.

36 MIZUHATA Shôzô, *Shin Imanishi shinkaron* (Evolution. For a new Imanishism), Nagoya, Chikyûshakai kikin, 2002.

37 Jacques DEWITTE, *La manifestation de soi. Éléments d'une critique philosophique de l'utilitarisme* (The manifestation of self. Elements of a philosophical critique of utilitarianism), Paris, La Découverte, 2010. Dewitte here follows Adolf PORTMAN, *La forme animale*, Paris, La Bibliothèque, 2013 (*Die Tiergestalt*, 2nd ed. 1960), on whose back cover he writes: "We find animal expressivity. Animals not only exist objectively, they show themselves to each other, they appear which is a fundamental function of the living." In the preface, he also writes on p. 20: "We must resign ourselves to translating '*die Gestalt*' by 'the form', even though the term means much more: not the form understood as a simple contour, but something like character, a way of being, a style."

38 *Op. cit.*, back cover.

39 *Op. cit.*, p. 174. This idea can be found both in the *Laozi* and more explicitly in the *Zhuangzi*. In the latter, the last sentence of chap. IV ("The human world", *Renjian shi* 人間世) says: "People all know the usefulness of usefulness, but they do not know the usefulness of uselessness" (*wuyong zhi yong* 無用之用).

40 One could also understand, at a lower level, society formed by the various species; in short, an inter-species sociality, in the framework of the geo-cosmos.

41 From Latin, *ad*, towards and *specio*, to look.

42 Used by François Vézin in his translation of *Être et temps* (Being and time), Paris, Gallimard, 1986; cf. the title of chap. III, "The worldliness of world", from the German *Die Weltlichkeit der Welt*. It is true that the German language distinguishes between *Weltlichkeit* and *Mondänität* (worldliness in the more frivolous sense), which is not what Heidegger meant. However, the social sciences, and particularly social history, lead me to believe that it is fundamentally the same thing. In *Histoire de l'habitat idéal*, op. cit. in § 7, I show how certain forms of worldliness in the leisure class appeared in the China of the Three Kingdoms and the Six Dynasties, and ended up 20 centuries later creating a life-style, urban sprawl, which not only made most of us cyborgs, incapable of living without a car, but is also upsetting the climate homeostasis of the planet. We are far removed from the early "worldliness".

43 Using again the concept in Thorstein VEBLEN, *Théorie de la classe de loisir*, Paris, Gallimard, 1970 (*The Theory of the Leisure Class*, 1899).

44 I leave the term in all its vagueness and slipperiness. On the one hand, the sociology of elites is never as established a field as the establishment would like it to be. On the other hand, the happy few of the leisure class do not hesitate to reinvent the customs of their inferiors as play, cf. Marie-Antoinette playing the shepherdess. Those games are not only symbols of the world, as Eugen Fink would have said, they are cosmo-genetic: they institute a world.

45 From *respecto*, repetitive use of *respicio*, to look backwards, to turn one's attention.

46 Cf. Georges VIGARELLO, *Le propre et le sale. L'hygiène du corps depuis le Moyen Âge*, Paris, Seuil, 1985, p. 105 (*Concepts of Cleanliness: Changing Attitudes in France since the Middle Ages*, Cambridge University Press, 2008):

> When he describes the bath taken in 1746 by the Marquise du Châtelet, Longchamp, then her valet, may surprise the modern reader; their relationship did not permit of modesty. The valet was responsible for the temperature of the water and supervised the boiler. From time to time he poured the contents into the tub, taking care not to burn the marquise. In short, he was there in the room, in attendance, attentive. The lady felt no embarrassment as she undressed and bathed, felt no inclination, even, to render the water opaque. The condition of mistress and valet were too far apart for decency to be threatened. The servant was not yet a person.

47 *Rôshi* (*Laozi*), 6, ed. by OGAWA Kanju, Tokyo, Chuôkôron, 1973, p.16.

48 I thank Dominique Buisset for correcting my pig Latin.

49 I draw here on Éric DARDEL, *L'histoire, science du concret* (History, science of the concrete), Paris, PUF, 1945.

Conclusion

Earth, to be sure, is the name we give it, but Earth is the one that asserts us

To summarize in one sentence this book's argument, let me appropriate, with all due respect, proposition 3.221 from Wittgenstein's *Tractatus Logico-philosophicus*, which says: "I can only *name* objects. Signs represent them. I can only speak of them, I cannot *assert them.* A proposition can only say *how* a thing is, not *what* it is" (*Die Gegenstände kann ich nur* nennen. *Zeichen vertreten sie. Ich kann nur* von *ihnen sprechen*, sie aussprechen *kann ich nicht. Ein Satz kann nur sagen*, wie *ein Ding ist, nicht* was *es ist*).[1]

Here we have in Wittgenstein's more colorful langage,[2] the meaning of the 100,000 or so preceding words: *Zwar nennen wir die Erde; aber sie ist* es, *die uns ausspricht.*

You know the translation already; it is the title above.

Nevertheless, let us gloss:

If human beings love the Earth; *if they really love it*, like the Weaver and the Herdsman who love each other so much that once a year, on the seventh night of the seventh month, they reunite above the Milky Way which had once separated them, then human beings must practice the tetralemma and especially the fourth one, the *syllemma*, the symbolic one that unites without confounding what the logos earlier disunited.

However, symbols are not sufficient. Between what things are and what we say about them, there is more than just words; there is what we are in our flesh, in our minds and in what we concretely do with them; in other words, in our mediance and its trajection. From the Earth, our foundation, up to the sky that attracts us.

The title above was indeed incomplete, too fated and ultimately deterministic. To be sure, the Earth asserts us, but we are the ones who say her: *zwar spricht sie uns aus, aber wir sind* es, *die sie sagen.* This is truly the throwing-beyond-oneself of a real trajection, of a real engendering, *genesis* …

And if we do not say what *it* means, the Earth, who will ever know? Lips that give birth to those who speak, forever. Not in some eternal return! But in the eternal advent of the one who, from herself, *ziran* 自然, is always to be born. *Natura natura semper!*

Notes

1 Wittgenstein's italics, p. 20 in the 2003 edition of the *Tractatus*, Frankfurt, Suhrkamp Verlag.
2 I thank Marion Picker for correcting my poor German.

Index

Page numbers followed by "n" indicate chapter note numbers.